For Marion & Ian

Fungi Lore

The Myths, Folklore and Magic of Mushrooms

Sally Coulthard

Illustrations by Clover Robin

Quadrille

Contents

Introduction

A few years ago, I built some raised vegetable beds using oak sleepers. Seasons have come and gone and the timbers, over time, have gently silvered and become sodden. These hunks of hardwood have also started to do something remarkable: they've begun to sprout masses of wild mushrooms. The grass around the base of the beds is dense with crowds of tiny Sulphur Tufts, pretty as a pixie-gathering and temptingly pickable. Around the sides of the timber boxes grow shelves of Oyster Mushrooms – huge, smoky grey, waxen caps with delicate white gills – and smaller numbers of the wonderfully descriptive Hairy Curtain Crust. Like a dandy's ruffled velvet sleeves, this handsome fungus gathers in flamboyant waves and is often gently pinstriped in oranges and browns. And, where the compost meets the timber, little clusters of mushrooms pop up that look like a baby's squashed ears – Palomino Cups, perhaps, or Blistered Cups, I'm not entirely sure.

And that, dear reader, is the nub of why mushrooms are so mesmerizing. Of those four species that have decided, uninvited, to colonize my garden, only one is delicious and wholesome. One is inedible but harmless. One is either inedible or toxic, depending on whether I've correctly identified it. And one is categorically poisonous. Get the choice right and you have a meal fit for a king. Get it wrong, however, and the results could be very unpleasant indeed.

And, surprisingly, it is the sweet, diminutively charming Sulphur Tuft that is the secret assassin. One mouthful and you'd be crippled with violent diarrhoea and vomiting before total collapse. Despite the latter's noxious secret, however, the Sulphur Tuft is also an ecological marvel. At dusk, this inconspicuous fungus starts to fluoresce under ultraviolet light, for a reason scientists still

haven't fathomed. And, as a wood-rotting species, the Sulphur Tuft also recycles nutrients back into the soil and is a true hero of the ecosystem. Truth is, if you value nature, it's hard to stay mad at a mushroom.

For thousands of years, people have been both drawn to fungi and rightly fearful of them; of the 150,000 species named around the world, as many are highly poisonous are as truly edible. Others take the human body to its mental and physical boundaries, inciting wild hallucinations or dangerous deliriums. Ancient legends speak of mushrooms as gifts from the gods, as tools for divination and as agents of change. Superstitions place them firmly in the world of witches and fairies. And folk wisdom has embraced the mushroom as a source of cure and sustenance, as well as death and destruction. Murderers' potions, symbols of prosperity, aphrodisiacs and even furniture for fairies – mushrooms have held different, curious meanings to different cultures. From the realms of medicine to the dark corners of superstition, fungi have shaped our ancestors' lives in ways few other organisms have.

They also occupy a curious niche in the natural world. Many grow in secret, or spring up without warning overnight. Others seem to live forever, glow in the dark or grow where little else can thrive. Fungi also develop close, intertwined relationships with other plants. And their weird and wonderful shapes and sizes, from the ravishingly beautiful to the downright odd, have earned them gloriously evocative folk names and beliefs. Countryside dwellers had richly vivid imaginations when it came to the world of the fungus. Take the Elfin Saddle (see page 52), for instance – a mushroom thought to be ridden by capricious elves. Or Dead Man's Fingers (see page 100), a macabre hand-shaped fungus that reaches up, like a buried cadaver, from the forest floor. Many of these whimsical nicknames have disappeared into history or been lost to more sober scientific labelling – obsolete but outrageously inspired names such as Beelzebub's Cushion, Sickener's Sister, Wrinkle-Twig or Guilty Sprout.

This book is a celebration of some of the world's most curious, mystical and much-loved fungi. Meet the picture-book Fly Agaric (see page 34) with its red-and-white cap, which has long been associated with shamanic visions. Or the impudent Stinkhorn (see page 10), which emerges from the earth like a grotesque phallic spear, pungent and perplexing. Find the Tippler's Bane (see page 128), a mushroom that turns an innocent alcoholic drink into a lethal potion. Or Liberty's Cap (see page 122), named after the hat of a freed Roman slave. However, this is more than just a guide to some of the world's best-known mushrooms. Here I reveal how medicine, food, magic, storytelling and ancient customs grew from one of the most mysterious of organisms. From grisly sacrifices to saddles for sprites, and fairy rings to king killers, it all begins with the humble mushroom. Welcome to *Fungi Lore*, an inquisitive and satisfying forage into the mysterious world of fungus.

Common Stinkhorn

Phallus impudicus

If you go down into the woods today, you're sure of a big surprise. Or, at least, you will be if you hang around until dusk. As twilight settles, one of the most extraordinary mushrooms will emerge with improbable vigour, turgid and phallic – for one night only – before it softens and crumples, its job done until the same time next year.

It is, of course, the Common Stinkhorn. Possibly the most tittersome and uncouth of all the fungi, it is a mushroom whose name perfectly describes its shape and challenging odour. Its survival strategy is improbably rude; to reproduce, the mushroom appears – like a tumescence – from the forest floor. Its looks are alarming for those of a delicate disposition, but nothing compared to its abominably putrid perfume – an aroma best described as a combination of dead cat and blocked drains. For just one evening of fetid glory, the stinkhorn will waft its scent through the undergrowth, hoping to attract carrion-loving insects, especially blowflies. These dead-flesh devourers will then clamber onto the cap of the stinkhorn, which is covered in a sticky spore-rich slime, before buzzing off to another damp, fertile corner of the forest.

In fact, the Common Stinkhorn lives up to its name so entirely that you are more likely to smell one before you see it. Its preference for rotting wood and other decaying organic material make it a relatively common dweller of woodland duff, but gardeners are often astonished to stumble upon a stinkhorn in their own backyards. Compost heaps, mulched flower beds, decomposing wood chip – a stinkhorn can make a fleeting appearance even in the politest suburban plots.

Over the centuries, we've come up with some gratifyingly rude names for the mushroom. John Gerard, in the 1597 first edition of *The Herball*, or the *General Historie of Plantes*, gave the Stinkhorn two immodest monikers. He called it *Fungus Virilis penis aerecti* – no translation needed – or the 'English pricke' mushroom, a name 'taken from his forme'. John Parkinson, warming to the theme in his *Theatrum Botanicum* four decades later, called it the 'dogges prike'. By the mid 18th century, however, the scientific world had settled on *Phallus impudicus*, or 'Shameless Penis' – a title that sounds like the sort of thing a prudish Georgian botanist might have muttered under his breath while hastily collecting samples for his botanicum.

For rural folk, however, the fungus earned other appellations. While some focused on its unholy pong – such as the 'Devil's Stinkpot', or 'Stinking Polecat' – others called it the 'Wood-witch', 'Troll's Egg' or 'Hedge-witch', its foul smell and uncanny, rapid surfacing blamed on malevolent forces. Some, nevertheless, looked to the phallic-shaped mushroom for magical potency. As a foolproof aphrodisiac, folk medicine prescribed collecting stinkhorn mushrooms, drying them over a fire or in the open air and then grinding them to a powder before adding a few teaspoons to a glass of something bracing. Such an enflaming concoction could also be used on farm animals to 'excite copulation'.

Not everyone reached for lewd uses of the stinkhorn. Writing the first pamphlet ever dedicated to this eye-opening fungus in 1564, Dutch physician Hadrianus Junius believed the stinkhorn to be 'a remarkable blessing for the wellbeing of mankind' and truly efficacious for 'intense and unbearable pains in the joints'. The fact that the mushroom looked uncannily like 'the male organ', he quipped cheerfully, was 'just proof of Nature's playfulness'.

If only everyone could have been so grown-up about it. Even the Darwins, a family well-versed in the peculiarities of the natural world, had their limits. Henrietta Darwin, daughter of Charles, is reported to have developed a deep antipathy toward *Phallus impudicus*. For her, the fungus's brazen appearance around the woodland was beyond her limits of tolerance. In a morally outraged horticultural purge, she was said to have dug up every offending stinkhorn she could get her hands on and tossed them on the drawing room fire.

Lobster Mushroom

Hypomyces lactifluorum

Imagine a nightmare world where a living organism is hijacked by a parasite, its metabolism rewritten, its colour and texture altered, and even its taste transformed. Welcome to the far-fetched and fascinating realm of *Hypomyces lactifluorum*, a mould-fungus that preys on forest fungi and turns them into something entirely new. What begins as a plain white mushroom quickly becomes a lurid, fiery-orange mutant – a striking anomaly in the forest undergrowth.

Mostly found in North America, this mould is picky in its choice of host, often targeting the Stubby Brittlegill (*Russula brevipes*) or Blancaccio (*Lactifluus piperatus*). The infection begins when the mushroom is dusted with the white spores of *Hypomyces lactifluorum*. From that moment, the transmogrification is swift and thorough – within just a handful of days, the host's delicate gills and pale flesh are overwhelmed. The fungus spreads both across the surface and deep into the mushroom, digesting tissues and replacing them with its own mycelial matrix. The result is a smooth, rigid rind, blazing in a sunset crimson, that completely masks the original species.

Yet this identity theft comes with a delicious reward. The parasitic mould-fungus reshapes the mushroom not just physically, but chemically: bitter compounds are neutralized, sugars and flavour compounds are concentrated, and the flesh becomes deliciously firm and dense – almost crunchy – even after cooking. It also creates the most fabulous shrimp-coloured natural dye. And the final surprise? Despite having absolutely no connection to crustaceans, the Lobster Mushroom delivers an unmistakable oceanic twist. Its aroma and flavour evoke cooked lobster or crab, turning an ordinary forest fungus into a culinary treat – a mushroom that not only looks like a lobster, but astonishingly tastes like one too.

Matsutake

Tricholoma matsutake

Some mushrooms, it turns out, are good at relationships. In nature, there is something called a mycorrhizal association – a partnership between a fungus that lives in the soil and the roots of a tree. The fungus and the roots work together and become biologically intertwined, each giving the other something it needs. The fungus gathers water and nutrients from the soil and shares them with its partner's roots. The tree, in return, offers the fungus something it cannot make itself: energy-rich sugar. It's a match made in heaven.

The science behind this arboreal affair was discovered only relatively recently, yet for thousands of years people noticed that certain fungus and plants were important to each other. Of all the mushrooms that enjoy this cosy connection, none is more famous or more prized than the Matsutake. Native to the forests of East Asia – especially Japan, Korea and China – the Matsutake has a unique bond with the roots of the red pine tree. Folklore describes the pine and the fungus as lovers, saying one cannot exist without the other. If the red pine is harmed by man, 'the Matsutake weeps' and disappears, a poetic interpretation of their real-life obligate relationship.

Matsutake mushrooms also have a distinctive fragrance. Their scent is often described as resinous, even cinnamon-like, and has become synonymous with autumn and its quiet repose. The *Man'yō-shū* (Collection of Ten Thousand Leaves), an 8th century anthology of Japanese poetry, includes a description of the earthy perfume of the Matsutake: 'On the peak of Takamatsu/so crowded you'd think there's no more room, mushrooms stand with their caps erect/ abundant and full; how wonderful is the fragrance of autumn.' This aroma was not only a sign of the forest's seasonal bounty but also a scent thought to attract *kami*, nature spirits in the traditional Shinto religion of Japan.

Turkey Tail

Trametes versicolor

Some fungi are gloriously cosmopolitan, popping up almost everywhere in the world. One such globetrotter is the Turkey Tail mushroom, a familiar sight on fallen logs in temperate forests across Europe, Asia and North America. Its name is well observed – this bracket fungus really does resemble the fanned tail feathers of a strutting turkey – and its handsome, contrasting stripes are as alluring to a fungi forager as an elaborate mating display is to a turkey hen. And, like any well-behaved flock, Turkey Tails tend to stick together, growing in sociable clusters on dead or dying hardwood trees.

While most Turkey Tails are resplendent in stripes of chocolate brown, clotted cream and soft grey, under certain conditions they reveal truly scrumptious rainbow hues – flashes of blue, rusty red and pale yellows. Despite their good looks, though, this leathery bracket fungus doesn't make good eating. Historic foraging guides suggest three days of boiling to render Turkey Tail vaguely edible, a culinary faff that does not repay the effort. Yet the mushroom is highly prized for its purported medicinal benefits, especially in China and Japan where it's known respectively as *Yunzhi*, the cloud mushroom, and *Kawaratake*, the roof-tile fungus. For centuries it has been dried, powdered or brewed into tea and prescribed for everything from colon cancer and lung infections to liver troubles and bronchitis.

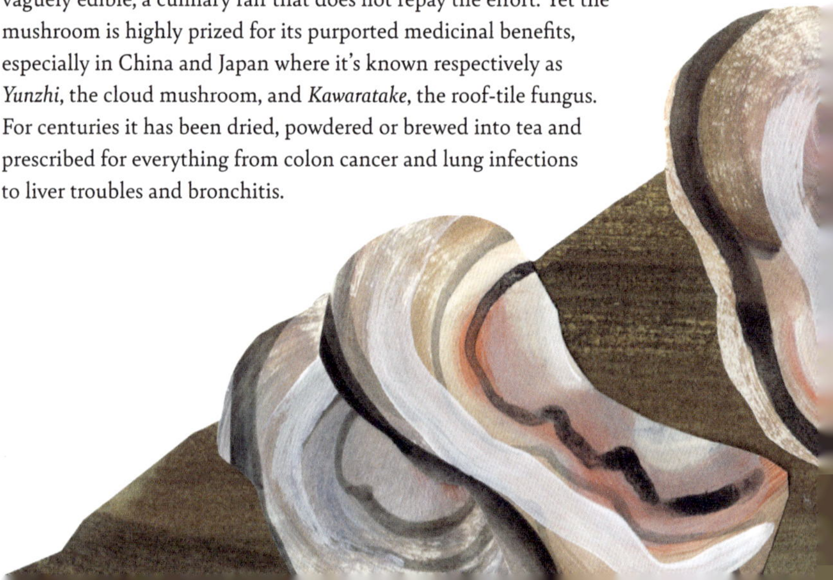

Once picked, Turkey Tail mushrooms stay wonderfully rigid, holding their shape and colour for months. This decorative quality made them popular in dried flower arrangements, posies and even as fashionable hat flourishes. But their true beauty lies in their work as quiet bioengineers, powering the cycles of decay and renewal that keep woodlands alive. Master decomposers, they break down the tough lignin and cellulose of hardwoods, returning nutrients to the forest floor and nourishing the next generation of plants, trees and microorganisms.

Giant Puffball

Calvatia gigantea

Thanks to the design of modern hives, beekeepers can harvest honey without disturbing their bees too much. Before the 19th century, however, collecting the amber nectar was slightly trickier. Many people kept their bees in woven skeps – bell-shaped straw or reed baskets – and collecting the honey involved dunking the entire hive into water to drown the colony, or killing the bees with poisonous fumes. There was, however, a third – less destructive – option for anyone willing to try it: smoke from a Giant Puffball.

Writing in 1783, Bryan I'Anson explained the cunning trick in his book *The Experienced Bee-keeper*. When a large puffball mushroom begins to turn brown, he noted, collect it and wrap it in brown paper. Squash the packet flat and place it in an oven where bread has just finished baking. Leave the parcel in the cooling oven overnight so that the puffball dries out. When you're ready to use it, cut off a piece as 'big as a hen's egg', set it on fire and pop it into the hive. In no time, the 'stupefying fumes' will render the bees completely insensible and you can grab the honey with impunity. By the next morning, the effects of the puffball will have worn off and the bees will be as lively as ever.

The Giant Puffball mushroom was, in fact, an enormously practical and useful mushroom for our ancestors. Writing almost seven decades earlier than I'Anson in his 1717 work *Lexicon Physico-medicum*, apothecarist John Quincy described the gargantuan fungus as 'the most powerful vegetable styptic yet known' and 'softer and more absorbent than lint'. For surgical operations and wound dressing, strips of Giant Puffball were applied externally to staunch bleeding, a practice common across temperate Europe and North America. Nineteenth-century doctors' reports record wadges of puffball being used regularly for uterine haemorrhages, deep lacerations, nose bleeds and bleeding after tooth extraction.

Inspired by the well-known effects of Giant Puffball smoke on bees, in the 1850s British physician Dr Benjamin Ward Richardson experimented with puffball fumes as an anaesthesia. A humane man who spent much of his academic life in the pursuit of pain-relief, for both people and livestock, between 1853 and 1860 he successfully anaesthetized more than a thousand animals and even practised the technique on himself, inhaling the fumes through a hookah pipe before passing out.

The Giant Puffball also made good eating. *St Nicholas* magazine, a popular monthly read for young Americans in the late 19th century, described the epicurean pleasures of the Giant Puffball: 'while they are young, firm and pure white [...] these large puff balls are very good eating sliced, seasoned well and fried in butter and especially if dipped in egg and cracker crumbs.' Its edibility is fleeting, however, as the Giant Puffball soon rapidly falls into decay, losing its pristine whiteness and breaking down into a dusty mass. In fact, almost everything about the puffball's life is transitory. As the children's magazine also noted, these curious growths 'sometimes appear as large as one's fist or larger in the morning in places where there was nothing of the sort the night before.'

John Parkinson, in his *Theatrum Botanicum* (1640), colourfully described the rotund fungus as having 'the biggnesse of a ball or balloone or a child's head more or less' and recalled a cheeky story of young lads throwing puffballs 'in sport' at one another. The original Latin name for the Giant Puffball – *Lycoperdon giganteum* – also dove into schoolboy humour; *Lycoperdon*, meant wolf's fart, no doubt a reference to the cloudy guff of spores emitted when a mature puffball bursts. It's an unusually playful scientific moniker and historic folk names across Europe also referenced breaking wind, albeit blaming different creatures of the forest. In Frisia it was the owl's fart, in Catalan the ox fart. In England it was the imp's, bull's or puck's fart, in Basque the ass's fart and in Sweden the old woman's fart. Some folk names ignored the puffball's flatulence – in certain English regional dialects it was the Devil's Snuff Box, the Fusball or Fumy Ball. Its spores were also believed to damage the eyes, a quality that earned the puffball its other nicknames: the Blindman's Ball or the Blindman's Bellows.

Lingzhi
Ganoderma sichuanense

Around 7,000 years ago, a society flourished on the coast of China. Archaeologists call it the Hemudu culture and, from the articles these ancient people left behind, it appears to have been a civilization that enjoyed rich and fulfilling lives. Not only were they among the earliest peoples to cultivate rice, the Hemudu also domesticated pigs, grew delicious crops and fished the surrounding waterways. And when they weren't finding food, they played musical instruments, crafted beautiful pottery and lacquerware, wove fabric and carved intricate jade ornaments. Their settlements reveal communities that blended practicality with artistry, suggesting a deep appreciation for both survival and beauty.

For a region considered the cradle of ancient Chinese culture – and one that produced such highly developed craft traditions – it's striking that Hemudu burials contained very few grave goods. The treasured objects that were occasionally placed with their dead, therefore, take on particular significance. Among the most intriguing of these were remnants of a very special mushroom: *Ganoderma sichuanense*. Its Chinese name, *Lingzhi*, has been translated as 'spirit mushroom' or 'mushroom of immortality', and in Chinese art often symbolizes wellness, great fortune and exceptional longevity. Rare in the wild, *Lingzhi* was historically accessible only to nobility, spiritual leaders and individuals of elevated status.

Over time, Lingzhi's reputation as the giver of vitality only grew. Today it is commercially cultivated and widely used in traditional Asian medicine for its purported immune-boosting, disease-resisting and anti-inflammatory properties. Given its long association with wealth, power and immortality, it's easy to see why the Hemudu may have chosen this fungus for their funerals. In honouring their dead with *Lingzhi*, it's tempting to imagine they were offering the ideal gift for a loved one on the threshold, stepping toward the afterlife.

Sky Blue
Entoloma hochstetteri

Names often reveal hidden stories, and few organisms wear theirs as vividly as *Entoloma hochstetteri*. This colour-saturated mushroom, native to New Zealand, is known by several evocative titles. Commonly called the Sky Blue mushroom or Blue Pinkgill, its names foreground the feature that makes it truly unforgettable: its colour. The fungus is an electric, crayon-like blue, an intensity so rare among mushrooms that it borders on the surreal. True blues are uncommon in the field of mushrooms – and the natural world at large – and this species displays the hue with unapologetic brilliance. Its pigment has even drawn the attention of chemists, who are investigating whether its colour can be isolated and reproduced, potentially unlocking a long-sought goal in food science: a truly natural, edible blue dye.

Scientific names, too, carry meaning. Before the mushroom's genus was renamed to *Entoloma*, the Sky Blue was long known as *Cortinarius* from the Latin *cortina*, meaning curtain or net. When young, the mushroom wears a fragile veil: fine, threadlike fibres stretching between cap and stem. This fleeting structure gives the fungus an ethereal appearance, as though wrapped in cobwebs or dressed for a spectral wedding, before the veil dissolves with maturity.

The most poetic name for the Sky Blue mushroom, however, comes from Māori tradition. For the Māori it is the *Werewere Kōkako*, a reference to the kōkako, a forest bird endemic to New Zealand. The kōkako resembles a dark grey crow but is distinguished by flamboyant blue wattles (fleshy flaps of skin) at the base of its beak. According to the tradition of the Tūhoe people, whose ancestral lands lie along the mountainous spine of the North Island, the bird acquired its blue wattles by rubbing its face against this luminous fungus. The name *Werewere Kōkako* translates, rather wonderfully, to 'the kōkako's wattles'.

FUNCTIONAL FUNGI

In 1991, a body was discovered trapped in ice in the Ötztal Alps on the Austria-Italy border. The man was remarkably well preserved, along with many of the objects he had carried with him. When his remains were dated, archaeologists were amazed to find that the 'Iceman', or Ötzi as he became affectionately known, had lived and died around 3300 BCE. His possessions also revealed intimate insights into everyday life five thousand years ago.

Ötzi was carrying with him two types of mushrooms: two pieces of Birch Bracket attached to a leather strap and a large quantity of Conk in his leather pouch. It's rare to find such ancient samples of mushrooms, but what's really interesting is that these two fungi were not being carried as food. Birch Bracket has another common name: Razor Strop. For thousands of years, this woody, cork-like mushroom has been used to sharpen and polish knives. Boiled in lye and beaten into a flexible material, Birch Bracket was used to make 'mushroom leather', a material that was utilized for purses, hats and decorative embellishments.

The name of Ötzi's other mushroom – Tinder Conk – also gives away its practical use. Thin strips of Tinder Conk made excellent fire-lighting material. It can hold a slow, smouldering ember for hours, allowing people to transport fire from place to place. Ötzi's fungi remind us that mushrooms have long shaped human lives in ways far beyond the kitchen. Across cultures, different species yielded dyes for wool and cloth. The Scaly Hedgehog, for instance, stains fibres deep forest green, while the Jack-O'-Lantern (see page 106) lends a bruised, spectral violet.

Other reported uses of mushrooms across small-scale societies include dried pieces of mushroom employed as bait in animal traps, as material for jewellery, for pomanders, fumigators and smokers, fishing floats, bottle stoppers, pin cushions and even toys. In Scandinavia, going back as far as Viking times, balls for various games were made from Birch Bracket, sometimes boiled in ash lye to make them more elastic. During World War II, when imported rubber supplies dried up, Birch Bracket balls were once again brought into play. Some fungi could even serve several purposes at once. The Giant Puffball (see page 20), for instance, provided tinder, ink and even a compress for septic wounds or bloodied flesh. Its drifting smoke also lulled bees into stillness and acted, in early medicine, as a crude but effective anaesthetic.

Blue Foot

Psilocybe caerulipes

While certain species of mushroom are showstoppers, others quietly grow in the wings. Science and folk medicine often focused on the former, but there are plenty of previously ignored fungi beginning to reveal their secrets. One such mushroom is the Blue Foot, a little, slender mushroom with a caramel cap and pale stem.

In late summer and autumn, in the deciduous forests of eastern North America, this well-camouflaged species grows by itself or in small clusters. It likes nothing better than rotting hardwood – the logs, branches and leaf litter from native trees. Overlooked by both indigenous cultures and colonial settlers, the Blue Foot was also described only in passing detail by most 19th century mycologists. It was, it seems, not pretty, potent or palatable enough to bother with.

And yet, the Blue Foot possesses two interesting traits. The first is one that gave the mushroom its name; if you pick a Blue Foot and accidentally bruise its stem, it starts to turn from white to cerulean. The second trait is that the Blue Foot contains psychoactive properties. More than two hundred species of mushroom contain the chemicals psilocybin and psilocin, both of which cause the brain to produce hallucinations and altered states of consciousness. While these two powerful compounds were only isolated in the laboratory in the 1950s, the clues were already lying in plain sight. Mushrooms that bruise blue often contain psychoactive compounds, as the colour change comes from the hallucinogenic chemicals reacting with air. This correlation has led many thrill seekers to follow the mantra that all 'blue bruisers' are psychedelic. No so. Wolf down a Satan's Bolete, for example, and the only trip you'll be going on is to the hospital.

Devil's Tooth

Hydnellum peckii

In conifer forests, across America and parts of Europe, including Scotland, huge teeth appear among the pine needle litter. Like mammoth molars, blocky and pale, these fungi appear as if ripped from the mouths of giants and tossed to the woodland floor. And, just as you might expect from a dental horror, these scattered teeth appear to seep drops of glistening, viscous blood.

This is the Devil's Tooth, one of the world's most extraordinary-looking mushrooms. Most of its folk names revel in its butchered, dentine appearance – it is also the Hydnellum, the Red-Juice Tooth, the Bile Tooth or Bleeding Tooth fungus. In German it's the Cork Tooth, in Slovakian the Burning Antler, and in Dutch the Blooddrop Prickle Fungus. Look closer, however, and this strange mushroom is touchingly pretty, especially when young; underneath those alarming globules is a fungus that's kitten-soft to the touch, with a velvet skin and warm pale-pinkish hue. Some have even thought it delicious-looking – the Devil's Tooth's only affectionate folk name is Strawberries and Cream. Anyone hoping for pudding will be sorely disappointed, however; Devil's Tooth isn't poisonous but is said, by those foolish enough to be tempted, to taste unbearably peppery and acrid.

The reason for the Devil's Tooth blood-splattered appearance isn't fully understood. Some suspect it might be an attractant, a compound irresistible to insects that will then carry the fungus' spores to another part of the woodland. Others believe it might do the opposite and repel any creature curious enough to consider the mushroom a potential meal. Despite its ruby looks, the Devil's Tooth creates a soft, green-brown colourant popular with natural dyers. The most compelling feature of the fungus' 'blood', however, is that it contains anticoagulant properties. What a trick Mother Nature has played, to hide a medicine for blood-clotting inside something that appears drenched in the stuff.

Fly Agaric
Amanita muscaria

Consuming fungi to get a spiritual 'high' is as old as time itself, with some of the earliest depictions of ritual mushroom-use coming not from the classical world or medieval art, but Stone Age cave paintings. In Algeria, for example, the Tassili n'Ajjer caves revealed paintings of 'magic mushrooms' created as early as 7000 BCE. These images, and other evidence from different cultures and throughout history, reveal a world where certain psychoactive fungi were central to ceremonial experiences and cultural life, turning feasts and funerals, births and marriages into psychedelic, otherworldly experiences.

Writing in the early 1800s, William Pinnock described one such mushroom – the Fly Agaric – and its heady reputation. In his ambitiously titled magazine *The Guide to Knowledge*, he noted:

'Many inebriating potations are from funguses but more especially from the Fly Agaric. The Russians [...] and others of the frigid regions of Northern Asia delight to steal their brains with its narcotic juices [...] This fungus is taken in the form of a bolus but more usually it is steeped in the expressed juice of [bilberry] berries when it stupefies with all the fatal speed of strong wine. Under its first influence the poor fool who drinks becomes as sadly comical as a Christmas clown: if he wishes to step over a straw he takes a stride or a jump sufficient to clear the trunk of a tree; a talkative person cannot keep secrets or silence and one fond of music is perpetually singing.'*

Pinnock, in danger of making a Fly Agaric spree sound a bit too much like fun, ended his description on a note of caution: 'thus he flounders on from one absurdity to another till outraged nature sinks into a stupor of unconsciousness or, more awful, bursts into spasmodic agony: the burning stomach heaves forth the cursed potion and it is well if the life go not with it.'

*a small lump of mushroom

So potent were Fly Agaric's charms that some people are reported to have collected and imbibed the urine of revellers and ritual goers. Writing in 1730, Philipp Johann von Strahlenberg, a Swedish officer who spent over a decade in Siberia, spotted the curious habit:

'the Russians [...] carry thither a kind of mushrooms [...] when they make a feast, they pour water upon some of these mushrooms and boil them. They then drink the liquor, which intoxicates them: the poorer sort, on these occasions, post themselves round the huts of the rich, and watch the opportunity of the guests coming down to make water; and then hold a wooden bowl to receive the urine, which they drink off greedily, as having still some virtue of the mushroom in it, and by this way they also get drunk'.

As a means of experiencing euphoria, Fly Agaric – with its unpleasant and potentially lethal side effects – was just too risky for most people to try. That didn't stop the mushroom, however, being useful in other ways. Its name comes from its practical application as an insecticide. Indeed, such is its efficacy as a bug-killer, it has been in use since at least the 13th century. Albert Magnus in his 1256 work, *De Vegetabilibus*, first described the fungus: *'vocatur fungus muscarum, eo quod in lacte pulverisatus interficit muscas'*, meaning, 'It is called the fly fungus, because when powdered in milk it kills flies' – a technique that endured over the centuries. Household guides from the 1950s were still recommending chunks of Fly Agaric steeped in a saucer of milk as a lethal flytrap. It seems the safest way to enjoy this notorious mushroom was to let the insects take the trip instead.

Clouded Funnel

Clitocybe nebularis

During the Victorian era, groups of amateur mushroom enthusiasts sprang up like fairy rings after rainfall. Around the country, mycology societies spent hours in the pursuit of fungi hunting: collecting specimens to classify and draw with scientific rigour. Some even suggested growing mushrooms as a new type of gardening; by creating areas of rotting leaves or manured land, gardeners could raise a cluster of edible varieties for the table. And one of the most popular suggestions for eating was the Clouded Funnel fungus.

Given its growing characteristics, it's perhaps no wonder folk wanted it to be edible. The mushroom has a large, nimbostratus-grey cap – often reaching up to 20cm (8in) in diameter – and pops up in densely packed, easily pickable patches or large rings, making it a woodland forager's dream. It also appears at the back end of the year, well into December, providing rich pickings when the pantry starts to look bare.

Anna Maria Hussey, in her lavishly hand-painted *Illustrations of British Mycology* (1847), couldn't have praised it enough: 'Of all the mycological dainties we have induced various accomplished gastronomes to taste', she enthused, this 'has had the most votes in its favour.' With an odour not unlike fresh cheese, the mushroom was scrumptious fried 'in a small quantity of fresh butter till crisp and slightly browned, pepper and salt being sprinkled upon.' It sounds delicious and many other fungi-lovers and cooks also commended its edibility. Unfortunately, in practice, the Clouded Funnel doesn't always reward such recommendation. Around a fifth of those who eat this mushroom will experience severe gastric upsets, leading modern foraging guides to change its status from edible to potentially poisonous. The Clouded Funnel is no longer regarded a delicacy, yet it remains a handsome presence in winter woods, only feeding our curiosity instead.

FAIRY RINGS

Few sights must have puzzled our ancestors more than the sudden appearance of a fairy ring: a neat circle of mushrooms rising overnight, as if placed by an unseen hand. Historically, some people blamed witches dancing under the moonlight or the sweep of a dragon's tail. Others muttered about lightning strikes or forgotten structures hidden beneath the soil. Yet across much of Europe, the explanation often fell to the fairies – those quick, impish creatures believed to slip between woodland and meadow. In old folk tales, the fairies spent their nights leaping and dancing in wild reels or galloping tiny horses around the turf. By morning, only a circle of fresh fungi hinted at their nocturnal revelry. In Shakespeare's *The Tempest*, we find this belief fall from the lips of Prospero:

'You demi puppets, that
By moonshine do the green sour ringlets make
Whereof the ewe not bites and you whose pastime
Is to make mid-night mushrooms'

The notion wasn't only reserved for superstitious villagers and inventive authors. Henry More, a 17th-century English philosopher, theologian and fellow of Christ's College, Cambridge, pondered with all seriousness whether fairy rings might 'be the rendezvous of witches, or the dancing places of those little puppet sprites which they call elves or fairies.' These magical circles were also warnings as much as wonders. Step inside a fairy ring, folk believed, and misfortune would surely follow: victims might find themselves vanishing without trace, dancing to exhaustion, or tumbling through a portal into the otherworld. The truth, however, is no less enchanting and lies in the soundless industry of fungi. A single specimen – often the Scotch Bonnet or Field Mushroom – will send out fine underground threads, radiating like spokes from a wheel. Each year, new mushrooms appear at the growing edge of this circle. The larger the fairy ring, therefore, the more ancient it is.

Scarlet Elf Cap

Sarcoscypha coccinea

Botanists of days gone by were rarely given to sentimentality. Unless, that is, they were writing about the Scarlet Elf Cap, one of the forest's most bewitching fungi. In the otherwise sober *Scottish Cryptogamic Flora* (1825), Dr Robert Kaye Greville allowed himself a rare indulgence describing this tiny red chalice. It is 'really entitled to be called a splendid fungus, and in point of beauty to be placed first on a list,' he wrote. 'The interior of the cup is of the finest carmine, the outer surface white and tomentose*. Without much poetical exaggeration, this beautiful [specimen] seems to be clothed with a fur robe lined with the richest velvet. So striking indeed is its appearance that I have had specimens brought to me by individuals of the lower classes solely on account of the splendour of their colour.'

In English plant lore, this picturesque fungus was also thought to be used by fairy folk – alternative names for this splendid mushroom all reveal these imaginative applications: the Fairies' Bath, Fairies' Hat, Blood Cups, Dryad's Cup** and the Fairy-Purse. It was also boiled as a cure for whooping cough, giving the mushroom its other folk name: Whooping Cups. The idea of fairies utilizing fungi as equipment, drinking vessels or clothing is long established. One of the earliest literary references appeared in *Oberon's Feast*, a highly imaginative and whimsical 17th-century poem by Robert Herrick. In it, Herrick conjures a miniature banquet hosted by the fairy king Oberon, where the tiny guests dine on ants' eggs, cuckoo-spit and fattened earwigs – all served to great ceremony on top of a fungus. His line, 'A little mushroom table spread', went on to inform many later folk tales and children's stories; in fairyland, fungi had become part of the furniture.

*fuzzy or downy

** a wood nymph

Cauliflower Mushroom

Sparassis crispa

Nature can be cruel sometimes. Imagine creating a mushroom that is easy to identify, large and wholesome, while also utterly delicious. And then also deciding to make it irritatingly rare. Welcome to the Cauliflower Mushroom, a fungus occasionally found in conifer forests across the northern hemisphere, including Britain and Ireland.

Its Latin name means 'torn and curly', but most foragers call it by its vegetable doppelganger, the Cauliflower Mushroom or Wood Cauliflower, and eagerly seek it out. It grows at the base of pine trees, firs and larches, popping up during late summer and autumn. Considered a prize find, the Cauliflower Mushroom was once a delicacy across many northern European regions – fricasseed in butter and herbs or baked with cream, garlic and egg yolks.

While most Cauliflower Mushrooms grow to around the size of an orange or small melon, occasionally they can reach gigantic proportions. Victorian mushroom-collectors, ever competitive, wrote up vivid accounts of finding whoppers in the wild. In 1881, a giddy member of the Essex Field Club reported that he had seen 'such a specimen as had probably never before gladdened the eyes of any fungologist', and described one 25cm (10in) high, over 1m (3ft) in circumference, and so heavy that it had to be cut in half and dropped into two fish baskets to be taken away. 'As far as the writer knows such another specimen has never been seen or even dreamed of in tale or history,' he bubbled with enthusiasm. Fungi enthusiasts wouldn't have to wait long, however, and just five or six years later another Cauliflower Mushroom of 'cyclopean proportions' popped up in Epping Forest, east of London. At 45cm (17¾in) high, and half as big again as the previous record-breaker, this *Sparassis crispa* was almost certainly 'the largest individual of this species' British fungus hunters had seen.

Tree Lungwort

Lobaria pulmonaria

For centuries, physicians subscribed to a fascinating theory: if you found something growing in nature that resembled a body part, or the symptoms of an illness, it could be used as the cure. So, for example, walnuts were thought to help brain disease, and testicle-like figs dealt with male infertility. It is, of course, total bunkum but that didn't prevent the Tree Lungwort – a member of the lichen family – from being liberally prescribed. John Gerard's 1636 edition of *The Herball* gives modern readers an insight into 17th-century thinking, telling us that it grows on old oaks, beeches and other 'wild trees in dark and thick woods.'

Gerard adds that its name Latin name – *Pulmonaria* – comes from its resemblance to the bronchioles of lungs and, as such, was recommended for breathing problems and tuberculosis. 'It is reported that shepherds and certain horse-leeches [horse 'doctors'] give the powder hereof with salt unto their sheep and other cattle which be troubled with the cough, and be broken winded*. Tree Lungwort is much commended of the learned physicians of our time against the diseases of the lungs, especially for the inflammations and ulcers of the same, being brought into powder, and drunk with water.'

In practice, there has been no scientific proof that *Lobaria pulmonaria* can act against lung disease. What is a marvel, however, is that the Tree Lungwort is not just one organism, but three. As with all lichens, the main body of the lungwort is formed by a fungus – but two types of algae live, symbiotically, just beneath its surface. There, these hardworking companions help the fungus photosynthesize and grab nitrogen from the air, while the Tree Lungwort, in return, provides structure and protection; a functional menage-a-trois if there ever was one.

*Also known as 'heaves', a bronchial condition in horses

Psilocybe Mexicana

Psilocybe mexicana

In the early 20th century, anthropologists and chemists went in search of a kind of holy grail – not a religious one, but a fungal one. When the Spanish first arrived in Mexico in the early 1500s, they observed that the indigenous Aztec people consumed a special mushroom during ceremonies and feasts. This fungus, known as *teonanácatl* (meaning 'flesh or food of the gods'), appeared to possess extraordinary intoxicating powers.

One of the earliest and most detailed accounts of *teonanácatl* came from the Franciscan missionary Bernardino de Sahagún, writing later in the 16th century. He described the mushroom's striking psychological effects:

'The first thing which they ate at the gathering was small, black mushrooms [...] These are intoxicating and cause visions to be seen and even provoke sensuousness. They ate these before dawn, and they also drank chocolate before daylight. They ate these little mushrooms with honey, and when they began to be excited by them, they began to dance, some singing, others weeping, for they were already intoxicated by the mushrooms.'

Unlike many of his contemporaries and later writers, Sahagún was careful not to dismiss the use of *teonanácatl* as mere drunkenness. Its consumption served a deliberate ritual purpose: to induce visions that could later be interpreted for meaning. Some participants, he noted, entered a contemplative or meditative state rather than dancing or singing. Others experienced vivid visions of death, violence, wealth, punishment or transgression. Once the effects subsided, the revellers gathered to discuss what they had seen and what it might mean. Such accounts captured the imagination of scholars, but for centuries no one could agree on what plant or organism *teonanácat*l was. Several psychoactive plants and fungi were proposed, including the peyote cactus, the Fly Agaric and various other 'medicinal' mushrooms. None of these candidates, however, fully matched the historical descriptions of a *teonanácatl* episode. The effects of the psychoactive substance – intensely visual, colourful and emotionally charged – were markedly different from the dissociative or trance-like states associated with Fly Agaric. *Teonanácatl* wasn't just a quick high for the Aztec people of the past – its hallucinations were read as revelations and divine omens.

A breakthrough came in the 1950s. In 1955, passionate mycologists Valentina and Gordon Wasson travelled to Mexico to participate in a little-known, sacred mushroom ceremony in Oaxaca, guided by an indigenous healer. They published an account of their experience in *Life* magazine in 1957, bringing global attention to the use of mind-altering mushrooms in traditional cultures. The mushrooms eaten during the ceremony were also identified as species of *Psilocybe*, including *Psilocybe mexicana*. Soon afterwards, in 1958, Swiss chemist Dr Albert Hofmann successfully identified the mushrooms' active compounds: psilocybin and psilocin. Hofmann, not content with simply isolating the chemicals, also decided to test them on himself:

'Whether my eyes were closed or open, I saw only Mexican motifs and colours. When the doctor supervising the experiment bent over me to check my blood pressure, he was transformed into an Aztec priest, and I would not have been astonished if he had drawn an obsidian knife. [...] At the peak of the intoxication, about one-and-a-half hours after ingestion of the mushrooms, the rush of interior pictures, mostly changing in shape and colour, reached such an alarming degree that I feared I would be torn into this whirlpool of form and colour and would dissolve. After about six hours, the dream came to an end.'

Elfin Saddle

Helvella lacunosa

In the middle of the 19th century, a Yorkshire merchant named Michael Aislabie Denham spent over a decade collecting regional folklore. Belief in fairies was widespread, even as scientific knowledge expanded, and many seemingly inexplicable things were attributed to these wily sprites. In his book *The Denham Tracts*, our author listed numerous natural objects or phenomena that were ascribed to 'fairies, fays and elves'. Static electricity was known as 'fairy sparks', and glow-worms were 'fairy lanterns'. Fossilized sea urchins were thought to be 'fairy loaves', while will-o'-the-wisps (bioluminescent marsh gases) were 'elf-fire'.

In British and North American folklore, certain mushrooms were also thought to belong to the fairies, including Giant Puffballs, Fly Agarics and Scotch Bonnets. Forest sprites were believed to ride various horses and woodland creatures, under the cover of twilight, with gossamer-thin webs for reins and mushrooms for saddles. Among these, the Elfin Saddle, or Black Saddle, is perhaps the most iconic. This charcoal-coloured, leathery-looking mushroom, common at woodland edges, has a deeply lobed cap forming a pommel and cantle, with a sunken central seat perfect for a fairy to straddle.

The appeal of fungi in fairy lore is no coincidence. For centuries fairy folk were thought to prefer liminal times, such as dusk and dawn, and thresholds between two places, such as crossroads, bridges and forest edges. Fungi fitted perfectly into this narrative world – uncanny growths that erupt spontaneously overnight often only to quickly vanish. Mushrooms also grow from decay, a state associated with the underworld, while their flesh can be both nourishing and noxious. Fungi including the Elfin Saddle, therefore, embodied the same capricious, unpredictable nature long ascribed to the fairy folk.

Collared Earthstar

Geastrum triplex

The Collared Earthstar looks extraterrestrial. When young, it resembles a tiny puffball, no bigger than a small, pale egg tucked into the leaf litter. With maturity, though, it performs a quiet transformation. The outer layer splits and peels back, cracking open like the hatch of an alien spacecraft, curling into pointed, star-like rays. Cradled within this opened shell sits a fragile inner sac, swollen with billions of spores and pierced at the tip by a minute hole. When rain strikes, wind passes or a footstep trembles the ground, a smoky puff erupts, releasing spore-stardust that drifts across the forest floor.

Despite its otherworldly appearance and peculiar unfurling behaviour, the Collared Earthstar is far from rare. It has been found on every continent except Antarctica and is just one member of a remarkably peculiar family. There are dozens of Earthstar species, each more improbable than the last. The Dark-Headed Earthstar looks as though it has been dipped in cocoa powder, while the Arched Earthstar lifts itself upright on its rigid 'arms', resembling a Martian tripod from HG Wells' *The War of the Worlds*.

Beyond their looks, Earthstars have long been woven into human culture. Ethnomycological records suggest that several indigenous cultures valued them as medicinal fungi, though exact species are notoriously difficult to identify. The Cherokee reportedly applied Earthstars to newborns' bellybuttons to prevent infection until the umbilical stump fell away. In Himalayan traditions, related fungi were used as salves for burns, while in Brazil the Rounded Earthstar – known as the 'Star of the Land' – was used to treat eye infections and asthma. One Earthstar served an entirely different purpose. The False Earthstar, a convincing lookalike rather than a true *Geastrum*, was once called the 'Poor Man's Weatherglass'. Its hygroscopic flesh absorbs moisture from the air, causing its starry arms to open and close with humidity levels – a living barometer, quietly blooming and withdrawing with the changing weather.

Chaga

Inonotus obliquus

Chaga belongs to the cold. Across large swathes of bitterly chill forest, from Siberia in the east to Canada in the west, this hard, charcoal-black fungus grows, like huge disfiguring carbuncles, on the trunks of birch trees. Its uses are anything but ugly, however, and it has provided native cultures with remedies and rituals for thousands of years. The Khanty people from Western Siberia used Chaga in their birthing rites. When a baby was born, it would be attended to by three midwives. The first, called the *pukhun-anki*, delivered the baby and cut the umbilical cord. The second, the *al'tem-anki*, purified the child in a bath of Chaga broth, and the last midwife, the *pema-anki*, chose the baby's name. Other ethnographic accounts of the Khanty document the use of Chaga to treat frostbite, intestinal worms, lung disease, gastric problems, liver disease and joint pain. Elsewhere, Chaga was made into an infusion or tea, inhaled as a snuff, or even made into an antiseptic body soap. In the Second World War, when Finland ran out of coffee, it turned to Chaga to make a drinkable substitute.

Set alight and allowed to glow as a hot ember, the Khanty also used lumps of Chaga to cauterize wounds, or to create an injury that triggered healing. A similar practice was observed in the Kamchadal people, over in Russia. Stepan Petrovich Krasheninnikov, an 18th-century explorer, described the medical procedure: 'For rheumatic pain they put a birch fungus on the injured places. When this fungus burns to the body it bursts out fiercely and the body decays and there is a great ulcer, on which some people pour fungus ash.' It was an extreme and highly risky treatment but modern medicine has not consigned Chaga to history; recent studies into the bioactive compounds of their remarkable fungus have shown it to hold some promising anticancer, anti-inflammatory and antimicrobial properties; a validation for those remote and rugged cultures that knew its value all along.

Earthball

Scleroderma citrinum

In Victorian fine dining, the truffle was king. And, as with all monarchies, there was fierce snobbery about lineage. The French worshipped their own Périgord Truffle, while the Italians swore allegiance to White Alba. Over the Channel, however, the English ate whatever they could get their hands on.

Britain's woodlands offered only four native truffles deemed edible, including the Summer and Autumn Truffles. That did little to deter 19th-century fungus hunters, who hauled to city markets anything remotely truffle-shaped. Alongside genuine finds came impostors such as King Alfred's Cakes, an inedible and tough briquette of a mushroom, and this, the Earthball. *The Encyclopaedia Britannica* (1894) described this 'rank and offensive' fungus as a 'false truffle', sometimes used to stuff turkeys and sausages in 'inferior dining-rooms of London where Continental dishes are served.' Spare a thought for diners unfortunate enough to be duped: Earthballs are poisonous, although thankfully not fatally, but consumption often led swiftly to vomiting and gastrointestinal distress.

The Gardeners' Chronicle had sounded the alarm four years earlier, noting the Earthball's increasing use in the adulteration of *pâté de foie gras*, as well as in supposedly 'truffled fowls' sold by less-than-conscientious butchers ('truffling' poultry involved slipping seasoned truffle butter between the skin and breast to keep the meat moist and infuse the bird with flavour). Victorian diners, it seems, were paying champagne prices for stomach-churning deceit.

Indeed, this unfortunate fungus never enjoyed kind press. Its nicknames included the 'warty potato', 'vegetable tripe' and 'poison pigskin puffball'. Even its Latin name, *Scleroderma* – meaning hard skin – was borrowed by medicine to describe an autoimmune condition marked by a thickened, toughened dermis.

And yet the Earthball is not entirely without admirers. One organism loves it above all others: the Parasitic Bolete (*Pseudoboletus parasiticus*), a mushroom that can survive only by clinging tightly to the Earthball and feeding from it. Proof, perhaps, that even the worst dinner guest is someone else's perfect host.

Old Man of the Woods

Strobilomyces strobilaceus

Almost everyone has heard of Beatrix Potter, creator of one of the most famous children's books of all time, *The Tale of Peter Rabbit*. Fewer people, however, know that she was absolutely mad about mushrooms. As befitted a Victorian woman of her class, Potter took a keen interest in natural history and found fungi especially captivating. Not only was she fascinated by their 'ephemeral fairy qualities' but, as an exceptionally talented botanical illustrator, the multifarious shapes, colours and textures of mushrooms proved a thrilling challenge.

In 1893, Potter spent a blissful summer away from London, holidaying in Birnam, Perthshire. There, she passed the time wandering woodlands and field

margins looking for specimens to paint. She was often accompanied by an elderly local mycologist called Charles McIntosh, who sported a long, wispy white beard and rimless spectacles, and spent much of his time on his hands and knees trying to get a better look at their finds.

In July that year, Potter stumbled upon a very rare fungus – Old Man of the Woods – a specimen that McIntosh had never seen in real life before. Less than a month later she found another and, inspired by their gentle beauty and pinecone-like texture, created three exquisitely detailed botanical paintings. The day after she painted the mushroom, she felt moved to write a letter to a young friend; a story about a naughty young rabbit called Peter, which went on to become the eponymous tale just a few years later. At the heart of the story Potter also placed Mr McGregor, an old man with a long, wispy white beard and rimless glasses, who begins the tale on his hands and knees in the garden. It is rather wonderful to think that Potter's mushroom hunting, and her friendship with McIntosh, inspired one of the greatest children's fictional characters of all time.

Death Cap

Amanita phalloides

Picture the scene: a grand country estate, a house party in full swing, laughter echoing through high-ceilinged rooms. The meal is sumptuous, the chat convivial until, suddenly, one guest begins gasping, clutching at his throat. Within minutes, he has collapsed onto the Persian rug, eyes frozen open in terror.

In the Golden Age of detective fiction, the art of the poisoner was indispensable, and many classic mysteries of the early 20th century feature deadly mushrooms as their instrument of choice. Beyond fiction, history too has hinted at mushroom-related fatalities, with the deaths of figures such as Emperor Claudius, Pope Clement VII and Charles VI linked to fatal fungi.

Perhaps the best known of all these 'killer mushrooms' is the Death Cap, a fungus so potently harmful that half a cap's worth is enough to slay a fully grown adult. It also murders in a blithely cruel fashion. Many reports describe the mushroom as particularly delicious and an unwitting victim will often feel well for hours after a hearty mushroom meal. Only perhaps, half a day later will the sufferer start to experience severe food poisoning-like symptoms before briefly rallying for a day or two. Don't be fooled, however, as the Death Cap is doing its work silently and methodically. There's a strong possibility that, before the week is out, their liver and kidneys will have failed and they'll have slipped into a deadly coma.

Of all the mushroom deaths worldwide, it's thought the Death Cap is behind nine out of ten. It's a fungus that's undoubtedly poisonous but, unlike some other equally noxious species, *Amanita phalloides* also wears the mask of edibility. When young, it looks like a number of forager-friendly fungi, including the Paddy Straw, Caesar's Mushroom and common Field Mushrooms, making it one of the most dangerous mistakes to make in the kitchen. And unlike some other mushrooms, its toxicity isn't touched by cooking, freezing or pickling.

Approximately a hundred people are killed every year from eating the Death Cap and yet not all *Amanita phalloides* poisonings are unintentional. One of the most notorious cases of death-by-mushroom involved a French man called Henri Girard.

In the early years of the 20th century Girard, a man who trained as both a chemist and insurance broker, hatched a plan to swindle his victims by taking life insurance policies out in their names (with himself as a beneficiary) and then poisoning them. He first experimented with bacterial cultures to give his targets typhoid, but after a number of unsuccessful attempts Girard turned to the steadfastly noxious Death Cap. *The Times* newspaper ran the shocking story in April 1921 with the headline: 'Fatal Luncheon Parties', and detailed Girard's ruse: 'When the insurances had been effected he invited each in turn to luncheon. After these meals the guests always became ill, and two of them, including the woman, died.' *The Kansas City Star*, later that same year, ran the even more sensational story following Girard's arrest: 'Slew with Mushrooms? End Comes in Prison to a Criminal of Imagination', which documented Girard's death while awaiting trial, remarkably by poisoning himself.

The story of 'Gentleman Girard' may have faded into obscurity, but the Death Cap's deadly potential persists. In 2025, the Leongatha mushroom murders in Australia brought it back to contemporary headlines. Erin Patterson was convicted of killing three relatives and seriously injuring a fourth by serving beef Wellingtons containing Death Caps at a family lunch. The case demonstrated how, even today, the fungus still has its place in the poisoner's handbook.

Jelly Ear

Auricularia auricula-judae

Tudor names for plants could often be vulgar; from poxwort to pissabed, open-arse to nipplewort, folk botanical names were sometimes shockingly rude. Yet the old name for the Jelly Ear mushroom – *Jew's Ear* – presents a particular conundrum. Some suggest it is a corruption of Judas' Ear, referencing Judas Iscariot, who, according to legend, hanged himself on an elder tree. But, as with many historical names, it might not be that straightforward.

The name appears in English texts as early as the 16th century. In *The Regiment of Life*, a translation of a work by French physician Jean Goeurot, we read:

'Take the musherun that groweth upon an elder tree, called in englysh, Jewes eares (for it is indede croncled and at, muche lyke an eare), heat it agaynst the fyer, and put it hote in anye drinke; the same drinke is good & holesome for the quinsy. Some hold opinion that whoso useth to drinke with it, shall never bee troubled with this disease, and therefore carye it about with them in jorneys.'*

It's easy to see why the fungus was associated with ears: the resemblance is uncanny. Steeped in beer, wine or vinegar, the mushroom was also thought to help deafness, either through drinking the concoction or applying it to the ear. But the Jewish reference is more complicated. Jews were expelled from England in 1290 and forbidden from re-entering for over 350 years. Yet cultural memory persisted. Antisemitic stereotypes – unfairly connecting Jews to treachery, poison and even dark magic – circulated widely in English literature, sermons and folklore. These prejudices sometimes shaped everyday language and naming conventions, including plant and fungi names.

Calling this fungus Jew's Ear was likely meant to hint at both its human-like appearance and the cultural suspicion of the unfamiliar. Today, we use the more neutral Jelly Ear, a name that celebrates its appearance and wonderfully boing-y texture. And leaves harmful stereotypes in the past, where they belong. Unfortunately, the Jelly Ear's Latin name is yet to catch up.

*A swollen, pus-filled pocket that forms near the tonsils.

Fool's Mushroom

Amanita verna

If you were trying to identify poisonous mushrooms in the early 1600s, John Parkinson's highly respected *Theatrum Botanicum* (The Theatre of Plants) would have been about as useful as a chocolate teapot. His chapter on 'Dangerous Mushromes' consisted of 32, horribly vague explanations of fungi known for making people ill. Descriptions such as that of the 'Toades Mushrome' epitomized Parkinson's want of any real-world knowledge on the subject; 'being of a wan colour, and round like a helmet, or more yellowish but smokey, or else more flat and reddish with some spots', could have referred to any number of mushrooms – some toxic, some tasty – and certainly wouldn't have given any forager a helpful steer. There were also some notable absences, including any recognizable mention of the deadliest of all fungi – the Fool's Mushroom or Spring Destroying Angel.

While the Death Cap mushroom *Amanita phalloides* holds the title for the most prolific killer, *Amanita verna* is thought to contain even higher concentrations of toxins per mouthful. Thankfully, its assassin-like reputation is redeemed by three notable differences: it is, thankfully, much rarer than the Death Cap; it appears in spring, not late summer or autumn (hence the folk name and the Latin epithet *verna*, or 'spring'), and it is so perfectly snow-white that it isn't often mistaken for an edible species. Before you breathe a sigh of relief, however, it is worth noting that there are lots of different varieties of Destroying Angels all around the world, including the Western North American Destroying Angel, the Eastern North American Destroying Angel, and the Guangzhou Destroying Angel. In Britain and Europe, alongside the Spring Destroying Angel, there is also its autumnal Destroying Angel cousin *Amanita virosa*, a mirror-image white mushroom that pops up in the second half of the year, a fresh invitation to make the same lethal mistake.

False Morel

Gyromitra esculenta

A few decades ago, in the French alpine village of Montchavin, a number of residents fell gravely ill with strikingly similar neurological symptoms. Medical researchers, confronted with an unusual cluster of cases, tested the drinking water, soil and even local radon levels in search of an explanation. After months of investigation, an unlikely culprit emerged: all of the patients had enjoyed local mushrooms known as the False Morel, which they picked in the mountain forests as the snow was melting.

For centuries, *Gyromitra esculenta* had been considered good eating, despite its rather brain-like caps. Its Latin name – *esculenta* – literally means 'edible'. Traditionally foraged and traded across Nordic countries, as well as parts of central and eastern Europe, it was enjoyed with much the same enthusiasm as the highly prized 'true morels', such as the Black and Common Morel. In Finland, its cultural status was such that it even appeared on a postage stamp in 1974. How, then, could the same mushroom be linked to devastating neurological disease in one community, yet seemingly leave others unharmed? The answer lies in gyromitrin, a naturally occurring toxin found in False Morels that affects the nervous system and damages DNA. Many northern cultures traditionally reduced the risk by repeatedly boiling and rinsing the mushrooms, or by drying them before consumption – processes that lower, but do not eliminate, gyromitrin levels. There is no fail-safe method, however, and repeated low-dose exposure can lead to cumulative poisoning.

Compounding the danger is the mushroom's fickle unpredictability. Gyromitrin levels seem to vary widely depending on soil chemistry, climate and local ecology – one man's meal is another man's poison. More troubling still, individual susceptibility differs markedly: some people are far more vulnerable to its effects than others. Given this uncertainty, it is perhaps unsurprising that several countries have chosen to ban the sale of this chancy lookalike altogether and consigned the *Gyromitra esculenta* to culinary history. How incongruous that a mushroom that looks so like the human brain could also be responsible for its damage.

Shaggy Inkcap

Coprinus comatus

Before the development of commercial inks in the 19th century, many people made their own. Ink could be created from a number of different materials including soot, charcoal, tree bark and oak galls – but one of the easiest methods involved the Shaggy Inkcap. Just as the fungus began to blacken with age and the gills dissolved, it was collected, mashed in a bowl and the juice strained. To stop the black ink from going mouldy, a few drops of highly toxic mercuric chloride would be added; or, failing that, a tot of vinegar in the hope that it would also prevent the ink from fading on the paper.

Such a useful mushroom rightly earned its folk name: the Shaggy Inkcap or Maned Ink-Cap, but this rather beautiful fungus also had many other monikers. For some it was the Goose Egg or Horse-tail Mushroom, the latter arising from its tufted surface. Indeed, another name for the Shaggy Inkcap was the Shaggy Mane. Perhaps the most evocative of all these folk names, however, was the Lawyer's Wig, a name that sprang up in the early years of the 18th century when wigs became part of formal court dress for barristers and judges. The Shaggy Inkcap's curled hairpiece also inspired another nickname: the Wig Dung-Stool, a slightly unfortunate title that described where the mushroom was often found. Indeed, the Latin name for the Shaggy Inkcap – *Coprinus comatus* – literally translates as 'dungy long hair'.

Those who focused on its ink, however, were missing the mushroom's real hidden treasure. It is, in fact, a delicious and highly nutritious mushroom when eaten young, a quality that has given it a much more appetising folk name in China. There, it is not known as the dung-stool or named after a preposterous hairpiece; it is the altogether more appetising 'chicken leg mushroom'.

Brick Top
Hypholoma lateritium

It's fascinating that, across the world, the same mushroom can be either ignored or eagerly foraged. One such fungus is the Brick Top (also known as Cinnamon Cap, Brick Cap, Brickie and Red Woodlover), a mushroom that most European countries leave well alone. In North America, however, it is collected and eaten with relish.

Writing at the turn of the 20th century, American professor and nature lover Dallas Lore Sharp wrote an entire chapter on the Brick Top in his book, *Wild Life Near Home*. He was amazed that so few people knew the true culinary delights of 'that October tit-bit, the brick-top mushroom'. It not only tasted delicious when correctly cooked or dried, but also grew in easily harvested clumps on old stumps. The *Hypholoma lateritium* was also straightforwardly recognizable and unlike anything else that might be collected by mistake. 'If it chances to be your first meal of brick-tops,' Sharp wrote, 'there is a stirring of the soul utterly impossible in the eating of a prosaic potato. You are on the verge all the time of discovery – of quail on toast, oysters, beefsteak, macaroni, caviar or liver, according to your nationality, native fancy and mycological intensity'. High praise indeed.

Fungus foragers in America, particularly along the east coast of America and throughout the Midwest, felt no fear of the Brick Top and frequently used it in stews, pickles or mushroom ketchup. The Japanese likewise enjoyed this dusky red mushroom after boiling it, calling it *kuritake*, or the Chestnut Fungus, after the chunks of rotten tree it grew upon. In Europe, however, mushroom hunters required far more convincing. Before the scientific world settled on *Hypholoma lateritium* – roughly translating as, thread-like, red-brick fungus, – the Brick Top was also known as *Hypholoma perplexum*. For many, therefore, it was the 'Perplexing Mushroom', a name that showed just how confusing fungus foraging can be.

Porcino
Boletus edulis

It's incredibly difficult to know what our ancestors ate thousands of years ago. Most plant and animal remains don't stand the test of time. But, just occasionally, you get a glimpse. The 'Red Lady' is a name given by archaeologists to the skeleton of a woman found in Cantabria in Spain. Researchers discovered at least two different types of mushroom spores trapped in the plaque in her teeth; one was some unknown variety of gilled mushroom and the other was a bolete. And, at an astonishing 18,700 years old, it's the oldest direct evidence we have that humans were eating mushrooms at this time – and not just accidentally.

There are a number of different types of boletes, but the most widely eaten is *Boletus edulis* or, as they are better known, Porcini mushrooms or Ceps. For a fungus we mostly associate with Mediterranean cuisine, the edible bolete is about as cosmopolitan a mushroom as you could wish for and can be found in temperate forests across the northern hemisphere and, as an introduced species, in Australia, New Zealand, South Africa and Zimbabwe. It is, however, a cornerstone of Italian cuisine, where it has enjoyed a long, if not always illustrious career. Pliny the Elder, writing in the 1st century CE, distinguished between high-quality edible mushrooms and Porcini, the latter of which he also called *suillos*, or 'pig mushrooms' – a dig, perhaps, at their lowly status. Even their modern name – Porcini – means little pigs. Snobbery aside, however, the Romans cooked freely with boletes and even named a dedicated cooking pot after them: the *boletari*.

Centuries later, European immigrants carried their mushroom-foraging traditions to North America. Not every old favourite grew in these unfamiliar forests but, happily, Porcini did. Better still, America even offered a supersized version: *Boletus edulis var. grandedulis*, the California King Bolete – a mushroom so big just one could feed a family of four. Think of it as taste of home, only bigger.

Morel
Morchella esculenta

In around 115 BCE, a law was enacted in Rome to try to curb some of the elite's worst excesses. Lavish fashion, sumptuous banquets, marathon drinking parties – it was all getting a bit embarrassing, and so the *Lex Aemilia sumptuaria* was passed. Among other things, the ruling prohibited certain kinds of foods at banquets; those seen as indulgent fripperies, including stuffed dormice and exotic birds. But as food historians note, when one door closes, another one opens. Denied their edible decadence, wealthy Romans turned to wild fungi instead, including Caesar's mushroom, Porcini, and the elusive honeycombed Morel.

The oldest collection of recipes to survive from antiquity, *De Re Coquinaria (The Art of Cooking)*, is attributed to Marcus Gavius Apicius (14–37 CE), a man who had fallen for mushrooms' charms. In the 'Sumptuous Dishes' chapter of his remarkably accessible cookbook he includes an entire section on mushrooms and reserves special mention for the magnificent Morel. He suggests three ways to please with this scrumptious specimen, including cooking Morels with garum [fish sauce] and pepper, Morels dressed in vinegar, wine and olive oil, and Morels boiled in salt water and dressed with fresh, chopped coriander. Simple, reverent and utterly delicious.

While Romans prized the Morel for its epicurean pleasures, other cultures leaned into its healing powers. In the Himalayas, where it's known as *guchhi*, indigenous communities – including the Bhotiya and residents of Himachal Pradesh's Kullu District – have long used Morels to treat gastric troubles, joint pain and even cuts and wounds. Local folklore insists the mushroom appears only when rain, lightning and thunder strike at once, as if summoned by heavenly drama. In Traditional Chinese Medicine, Morels are believed to reduce phlegm and boost immunity by regulating *qi*, the body's vital energy. In Mandarin, they're called *yang du jun* – Lamb's Stomach Mushroom – an offally unfortunate name for one of the most highly prized mushrooms on the market.

Purple Pinwheel

Marasmius haematocephalus

If a magical fairy needed an umbrella, they'd look no further than the Purple Pinwheel. It is one of the world's daintiest fungi – a tiny parasol painted in the prettiest of mauve pinks. And, as befitting a Lilliputian brolly, its canopy is miniscule – often no more than a centimetre across – with a fine, slender shaft in a darker hue and crisply defined spokes.

Just like an umbrella, the Purple Pinwheel excels in the rain. Its Latin name *Marasmius* comes from the ancient Greek word *marasmus*, meaning to wither or waste away, a name that hints at the Purple Pinwheel's most enchanting trick. When the forest's weather turns dry, the little fungus does not fight the drought. Instead, it folds itself down, shrivelling and dehydrating, until it seems barely alive; a trait known as marcescence. Yet this is no surrender. With the first returning rains, the Purple Pinwheel drinks deeply, swelling back into shape. While most mushrooms gamble everything on a short, fleeting period of damp weather, the Purple Pinwheel is patient. It can revive and release its spores repeatedly, waiting out the dry times and plumping itself back to life when the heavens reopen.

Mushrooms called 'Purple Pinwheels' can be found across the world's tropical and subtropical regions, but the true *Marasmius haematocephalus* is thought to occur only in South America. Its continental cousins look remarkably similar, however, so the next mystery for scientists is how such alike mushrooms ended up so widely distributed. One mesmerizing theory is that mushroom spores – which are microscopically small – can be swept high into the atmosphere and travel for hundreds of miles. When raindrops form around airborne particles, some of these spores fall back to earth miles from where they began, growing tiny little purple umbrellas wherever they land.

Witch's Butter

Tremella mesenterica

The Scottish preacher George Sinclair published *Satan's Invisible World Discovered* in 1685. It was a rigorously researched – and utterly sincere – attempt to prove that the Devil was real, active and best defeated through unwavering Christian faith. Satan, Sinclair argued, worked through many channels: witches, demons and a menagerie of shape-shifting creatures.

Sinclair called these animal familiars 'carriers' and retold the confession of an elderly Scottish woman who – no doubt under torture during the witch trails – had admitted to owning one:

'They confessed also, that the devil gives them a beast, about the shape and bigness of a cat, which they call a carrier; and he gives them a bird, too, as big as a raven, but white: And these creatures they can send anywhere and wherever they come, they take away all sorts of victuals they can get, as butter, cheese, milk, bacon, and all sorts of seeds, whatever they can find, and carry it to the witches. What the bird brings, they may keep for themselves: but what the carrier brings, they must reserve for the devil, and that is brought to Blockula, where he gives them of it as much as he thinks fit. They added, that the carriers filled themselves so full oftentimes, that they are forced to spew by the way, which spewing is found in several gardens, where colworts grow, and not far from the houses of the witches. It is of a yellow colour like gold, and is called the butter of the witches.'

The Witch's Butter mushroom Sinclair describes is now known as Tremella *mesenterica*: a glossy, brain-like jelly fungus, glowing dairy-gold on dead wood across much of the world. Though Sinclair writes from Scotland, his story points unmistakably east. He names *Blokula* or *Blåkulla*, the 'blue hill' of Swedish folklore: a mythical island where witches were believed to gather and consort with the Devil.

In Swedish – and broader Nordic – witch lore, the sudden appearance of Witch's Butter or the similarly lurid Scrambled Egg Slime fungus was bound up with anxieties over milk and butter; vital commodities in the peasant economy. Witches were thought to send familiars to steal dairy from domestic pantries and neighbouring farms. Any magical spillage on the journey home – too much stolen milk, for example, or a carrier who had drunk too much and vomited – was left behind as these suspicious yellow growths.

Because such thefts were understood as diabolical in origin, the fungus had to be destroyed properly. Accepted methods included burning it, beating it to shreds, whipping it with hawthorn (a tree heavy with magical associations), or consigning it to the fire at a crossroads.

At the end of the 18th century, the Swedish priest Johan Peter Wallensteen described one such ritual:

'Newly grown fungi on old fences or walls are called witch butter. It is believed that they are created by an evil person who wants to destroy the barn or churned milk. These fungi are whipped and beaten, as if in anger, while swearing and cursing the Devil [...] finally the people prepare a fire [...] scrape carefully with sticks whatever is left [...] disgusted to touch it with bare hands [...] and have fun watching how these fungi boil and burn in the fire, very happy about having discovered the evil and about the victory they have gained over it.'

What looks today like harmless fungus was once proof of theft, witchcraft and the Devil's appetite for dairy – indisputable evidence that something had gone badly, supernaturally wrong.

TO EAT OR NOT TO EAT?

In the history of mushrooms, it has long been something of a puzzle why many European countries developed rich traditions of fungal foraging while the British largely avoided it. The same species grow across the continent, yet for centuries there has been a marked difference in how fungi were used and eaten.

Several explanations have been offered. One is that the Romans introduced knowledge of edible mushrooms to Britain in the 1st century CE, only for that knowledge to be lost after their withdrawal three centuries later. Another suggests that Britain's abundant pasture and fertile farmland made foraging unnecessary. A more convincing explanation, however, lies in the landscape itself. Much of mainland Europe remains heavily forested – ideal environments for fungi – while Britain was largely deforested by the early Middle Ages. By the time William the Conqueror arrived in 1066, only around 15 per cent of Britain was wooded, compared with vast tracts of forest covering as much as half of mainland Europe.

Knowledge of edible fungi is built on generations of hard-won experience. Across Europe, food markets have long sold wild mushrooms gathered by people with an intimate understanding of different species. In some regions, a paid 'inspector of fungi' acted as a safeguard, examining foragers' baskets and checking sellers' stalls for dangerous specimens. Identification also relied not only on a mushroom's colour, smell or shape, but also on its habitat, seasonality, rarity, life cycle and whether it grew alone or in groups. This knowledge was communal, practical and inherited.

Britain, by contrast, developed a strategy of avoidance, or 'mycophobia'. Without forests to sustain regular foraging, and without a strong tradition to refine identification, mushrooms came to be seen as unpredictable and dangerous. Poisonings reinforced fear rather than understanding. In other words, Europe learned which mushrooms to fear while Britain learned to fear mushrooms.

Termite Mushroom

Termitomyces titanicus

There is an arrangement in nature that any smallholder would admire. Long before humans abandoned hunting and gathering in favour of agriculture, termites in Zambia were already farming. And, doing it very well indeed. Deep underground, inside their labyrinthine nests, termites cultivate a singularly strange crop: the Termite Mushroom. The insects ferry dead plants, dry grass and fragments of wood into their subterranean compost heaps, where the fungus gets to work. With a touch of biochemical magic, the fungus breaks down tough plant fibres that the termites cannot digest and, in the process, begins to grow. In return for this endless shovelling of organic matter, the termites harvest the fungus's mycelium, taking what they need while leaving the living system intact. It is a relationship built on balance and restraint – like picking apples from a tree rather than cutting the whole thing down.

When the rains arrive, the fungus sends up fruiting bodies of frankly eye-watering proportions. *Termitomyces titanicus* is the largest edible mushroom on Earth, its pale caps stretching up to 1m (3ft) across, heaving itself out of termite mounds like something from a fairytale. Unsurprisingly, local Zambian folklore imagines these vast mushroom caps as shelters for snakes and antelopes, a reminder that the fungus belongs as much to imagination as it does to ecology.

Among the Bemba people, mushroom collecting is not a casual pastime but a female-led knowledge system, passed down with care. Grandmothers and mothers teach girls how to read the weather, soil and seasons, and how to recognize abundance without exhausting it. Known locally as *ichikolowa*, the Termite Mushroom is happily both edible and excellent. With its dense, meaty texture and deep flavour, this colossus of the soil is a much-anticipated gift of the rainy season.

Basket Stinkhorn

Clathrus ruber

There are only a handful of mushrooms that were recorded by science in the 1500s. One of these lucky few was *Clathrus ruber*, a fungus so eye-catching it's perhaps no surprise it caught the attention of early natural historians. Published in 1560, Conrad Gessner's *Nomenclator aquatilium animantium* included two illustrations of this latticed beauty – also known as the Basket Stinkhorn – although he wasn't entirely sure what it was. His book, which provided one of the first truly detailed guides to aquatic animals, placed the mushroom in the 'De Crustatis', or 'hard-shelled invertebrates', chapter, alongside crabs, lobsters and shrimps. Gessner seems to have been convinced that, because of its net-like body and tough skin, the Basket Stinkhorn was almost certainly a creature of the sea, and he duly named it *Fungos marinos*.

You can hardly blame Gessner for the confusion. Not only does the Basket Stinkhorn look like nothing else that grows on land, but it also seems positively otherworldly. Emerging from a pale, egg-like structure, its grilled-fenestration expands rapidly, forming a hollow cage that appears divinely engineered rather than grown. Its Latin name means 'red trellis' or 'red cage,' a description that does exactly what it says on the tin, but its folk names are far more evocative. In Croatia it is *Vještičino srce*: Witch's Heart. In France, *Coeur de Sorcière*: Sorcerer's Heart. In Catalan, *Gita de Bruixa*: Witch's Egg, or *Reixa del diable*: Devil's Gate.

In several European countries, folk tradition also warned that touching the fungus could cause tumours, skin diseases or convulsions.

Adding to its unsettling reputation is its hideous smell. Like other stinkhorns, *Clathrus ruber* produces an odour reminiscent of rotting fish, designed to attract dead-flesh-loving flies that spread its spores. Seen through the eyes of a bemused 16th-century naturalist, it is little wonder that Gessner mistook the Basket Stinkhorn for something dredged from the sea, rather than an organism rooted in soil.

Octopus Stinkhorn

Clathrus archeri

As we saw with the Basket Stinkhorn, natural historians of the past could sometimes muddle their species. Medieval bestiaries – books cataloguing the plants and animals of the known world – freely blended observation with imagination. Alongside careful descriptions of real flora and fauna sat utterly fanciful creatures. Unicorns, griffons and dragons were as real to the medieval mind as the mouse, cow or dog. There was also a widespread belief that the sea mirrored the land. For every terrestrial horse, there was a seahorse; for every hedgehog, a sea urchin (urchin being the old word for hedgehog); and for every lion, of course, a sea lion.

Such ideas sound ridiculous until you come face to face with the Octopus Stinkhorn, *Clathrus archeri*, a fungus that looks as though it has crawled straight from the pages of *Twenty Thousand Leagues Under the Sea*. This cephalopod-mimic begins life as a fragile, golf-ball-sized 'egg'. When it is ready to release its spores, long, red, tentacle-like arms burst free and slowly arch backwards into a distinctive squid-like form – an emergence made even more unsettling if captured on time-lapse film.

Despite its kraken-ish appearance, the Octopus Stinkhorn is a gentle presence, quietly erupting from the woodland floor before collapsing back into a black, sea-bed sludge. In truth, the only thing it really shares with its marine counterpart is a talent for ocean travel. Native to Australia and New Zealand, mycologists believe its spores somehow managed to smuggle their way aboard ocean-going cargo – perhaps in wool sacks or other World War I supplies – eventually landing in Europe, where this bizarre organism sporadically rears its tentacles.

Satan's Bolete

Rubroboletus satanas

Harald Othmar Lenz's first encounter with the Satan's Bolete is meticulously recorded in his diary. On 12 September 1830, the German naturalist and his friend Karl Salzman, a medical student, set out for what promised to be a pleasant day of mushroom foraging. Though the weather was miserable and wet, it did little to dampen their spirits, and by the end of the day they returned with baskets 'filled with rich bounty'.

Among their finds was one unfamiliar mushroom that particularly caught Lenz's attention: a plump, meaty specimen with a red stalk, whose flesh turned blue when cut. This attractive-looking fungus was also generously sized and had a firm, enticing texture. Cautiously, Lenz nibbled a small piece, then spat it out – just to be safe.

Half an hour later, he described a sudden sensation 'shooting through his brain', followed by relentless and repeated bouts of vomiting. For three days Lenz endured weakness, sickness and diarrhoea before, finally, he could face the world again. That same evening, Salzman visited and casually mentioned that he had just enjoyed a lip-smacking fry-up of the very same mushroom, sautéed with butter, bacon and onions. Unsurprisingly, Salzman spent the next several days equally indisposed.

Once fully recovered, Lenz decided to claim this 'hitherto unknown' mushroom for science. He named it *Boletus Satanas* – Satan's Bolete – and declared it one of the most poisonous mushrooms he had ever encountered in his extensive mycological studies. Folk names from around Europe suggest that local foragers had reached similar conclusions long ago: in Italy, it is the *Porcino Malifico,* the Evil Fungus; in Catalonia, *Mataparent,* the Killer of Relatives or *Corvall del Dimoni,* Devil's Horn; in France's L'Occitane region it's *Camparòl fòl,* the Insanity Fungus; and in Poland *Krwistoborowik Szatański,* the Bloody Devil.

Diamond Willow

Haploporus odorus

During his 1732 tour of Lapland, the 'father of modern taxonomy' Carl Linnaeus took careful note of what local people were doing with fungi. He was particularly astonished to see local Sami men using mushrooms to make themselves irresistible:

'The Lapland youth, having found this Agaric, carefully preserves it in a little pocket hanging at his waist, that its grateful perfume may render him more acceptable to his favourite fair-one. O whimsical Venus! in other regions you must be treated with coffee and chocolate, preserves and sweetmeats, wines and dainties, jewels and pearls, gold and silver, silks and cosmetics, balls and assemblies, music and theatrical exhibitions: here you are satisfied with a little withered fungus!'

This fragrant 'agaric' was the *Haploporus odorus*, also known as the Diamond Willow Fungus, Willow Tick or Scented Bracket. It is a fungus of far-northern, old growth forests – in Sweden, Finland, Russia and parts of Canada and North America – where it grows on the trunks of gnarly willow trees. It's not much to look at – just a clump of stiff white marshmallow – but its scent is strongly reminiscent of old-fashioned candy. Think aniseed, vanilla and perhaps a hint of liquorice, floating out into the forest like the open doors of a sweet shop.

Once cut, this anise-like growth will keep its scent for months and, in Scandinavia, lumps of Diamond Willow were traditionally tucked into clothing chests as a moth repellent. There, it was also used to give an aniseed seasoning to porridge, cheese, bread and rice pudding. The fungus' long-lasting fragrance also encouraged some Northern Plain Native American tribes – including the Chipewyan and Cree – to use the mushroom as both a practical cure and a potent spiritual tool. Its smoke was central to smudging rituals, to cleanse and bless spaces, as well as communicating with ancestors.

Taken as an infusion, the fungus was prescribed for stomach problems and coughs, and its juice squeezed into the ear for earache. Perhaps the most interesting use of the fungus, however, was as an amulet of sacred power; nuggets of dried Diamond Willow were threaded onto necklaces or sewn onto clothing as candy-sweet protection from harm.

Chicken of the Woods

Laetiporus sulphureus

What does the mushroom *Laetiporus sulphureus* have in common with a hen? Well, quite a lot as it turns out. To begin with, both are remarkably ubiquitous. Just as the domestic chicken has travelled the globe alongside humans – becoming woven into cuisines and food traditions on every continent – so too has this striking bracket fungus. Known to foragers across Europe, Asia and the Americas, *Laetiporus sulphureus* has a long history of human consumption. Its bright sulphur-yellow shelves are hard to miss, even for the most inexperienced forager, and its value as food has been recognized for centuries.

Then there is the matter of taste and texture. When cooked, Chicken of the Woods has a fibrous, juicy mouthfeel and a mild flavour said to resemble poultry meat. This palatable similarity has earned the fungus a whole flock of gallus-inspired names, including Chicken Polypore or simply the Chicken Mushroom. The parallels even extend into genomics.

The chicken was the first bird to have its genome fully sequenced, becoming a foundational model organism in biology. In a similar spirit, *Laetiporus sulphureus* was selected as the first fungal genome sequenced within the Darwin Tree of Life project, an ambitious UK-led programme aiming to decode over 70,000 species of animals, plants and fungi across Britain and Ireland.

What draws science to both organisms is, in part, their enormous commercial potential. Chickens have underpinned the global protein economy since the mid-20th century, though often at significant environmental cost. With nutritional science and the pharmaceutical industry looking for more sustainable ways to feed a growing population, and keep it healthy, the Chicken of the Woods is looking cautiously promising. Dried, it is unexpectedly high in protein and low in fat – much like chicken meat – but with the bonus of a significant and welcome dose of dietary fibre. It is also a fungus rich in bioactive compounds, ones that might be able to offer anti-inflammatory, antimicrobial and even anti-malarial properties.

Dead Man's Fingers

Xylaria polymorpha

In the 12th century, a Catholic priest recorded a spine-chilling account of a dead adulterer who, after his burial, rose again from the soil. William of Newburgh's *History of English Affairs* pulled no punches in its description of the terrifying evening's events:

'by the handiwork of Satan, from his grave at night-time [...] he wandered through the courts and around the houses while all men made fast their doors, and did not dare to go abroad on any errand [...] until the sunrise, for fear of meeting and being beaten black and blue by this vagrant monster. But those precautions were of no avail; for the atmosphere, poisoned by the vagaries of this foul carcass, filled every house with disease and death by its pestiferous breath.'

This was no idle hearsay, but a report written in sober seriousness. For thousands of years, people across Europe believed in the idea of 'revenants': corpses that returned from the grave to torment the living. These were not rejuvenated bodies or romantic spirits, but walking remains – bloated, blackened and unmistakably dead. Those thought most likely to rise were rarely ordinary folk; they were sinners, social outcasts, practitioners of forbidden magic, or people who died 'unnatural' deaths, their souls imagined as unsettled and vengeful.

Revenants were said to emerge at liminal times – most often at dusk – haunting through the night until dawn's light forced them back to their graves. Fear of such resurrections shaped burial practices across medieval Europe. Corpses of potential revenants might be weighed down with stones, staked through the chest, or buried face-down. Coins were placed in mouths, limbs were bound, and in some cases heads were even removed entirely. The dead, it seemed, required firm supervision.

Given these anxieties, vigilance did not end with the grave. Communities scanned their natural surroundings for ominous signs, and few sights would have been more unsettling than Dead Man's Fingers. Emerging as if from the underworld, these macabre-looking mushrooms resemble a clump of human digits. They reach upwards, as if imploring, a pallid pale-grey hand of the recently deceased. As they mature, the resemblance only deepens: the surface darkens, wrinkles and hardens, evoking the preserved skin of an ancient cadaver. Little is known about early folk attitudes toward Dead Man's Fingers specifically, but the association between fungi and revenants persisted well into the 19th century. Percy Bysshe Shelley captured this unease vividly in verse:

'And agarics and fungi, with mildew and mould
Started like mist from the wet ground cold,
Pale, fleshy, as if the decayed dead
With a spirit of growth had been animated!'

Yet this macabre interpretation was far from universal. *Xylaria polymorpha* is found across the world, and in other cultures it inspired not fear, but respect. Among the Baiga and Bharia – forest-dwelling tribes of central India – the fungus was known as *Phoot Dooth* (meaning 'milk gusher'). There it was valued as a medicine used to encourage milk production in new mothers struggling to lactate; the dried fungus was ground into powder, mixed with sugar and taken with cow's milk for five consecutive days. In Ayurvedic medicine, it is similarly prized as a bringer of fluids but, in this instance, as a powerful diuretic. Meanwhile, among the Temuan people of Peninsular Malaysia, Dead Man's Fingers are used for almost the opposite purpose: to treat bedwetting in children.

Modern science has added yet another layer to this story. Far from being a herald of death, *Xylaria polymorpha* has been shown to possess potent antifungal compounds. Dead Man's Fingers are also, like other fungi, first-rate recyclers and transform dead, woody material into soft, biologically usable matter. What once appeared as the grasping hand of a revenant is, in fact, part of nature's vibrant circularity – breaking down old material so new life can take hold.

Oak Mazegill

Fomitopsis quercina

Viking York, or *Yorvik*, in the north of England has revealed many secrets over the decades. Excavations uncover a world rich in colour and activity, where people used the natural materials around them not only for food, but for a wide range of practical and medicinal purposes. And, among the debris from one busy quarter of the city was an intriguing find: a fragment of Oak Mazegill fungus or *Fomitopsis quercina*.

This piece was recovered from deep within a pit filled with refuse. Such waste holes often contain bones, vegetable scraps, human and dog faeces, and broken or discarded household items. But what was a bracket fungus doing in a Viking Age rubbish dump? To answer that question, archaeologists look beyond York, drawing on evidence from other sites, periods and cultures. Oak Mazegill, despite being inedible, has been remarkably useful across history. One of its most widespread uses was as tinder. Its tough, woody brackets are riddled with large, gill-like pores, which, when dried, can hold a glowing ember for hours. The same fungus has been found at some very ancient sites indeed, including La Draga, a 7,000-year-old lakeside settlement in Spain. There, researchers believe it formed part of a portable 'tinder kit' for carrying fire from place to place.

Oak Mazegill also had other down-to-earth uses. When dried, its maze-like surface made it an effective 'curry comb' for grooming horses flanks and hindquarters. Elsewhere it has been recorded as a sponge, a substitute for puffballs when smoking bees, or even as a wound dressing. Given the cess-pit context of the *Yorvik* find, however, another possibility is that it served as a form of toilet paper – soft organic materials such as moss were commonly used for personal hygiene. Whether this small fragment once lit fires, brushed horses, dressed wounds or performed a less glamorous task, we'll never know for sure. But even in the Viking world, it seems fungi could have quite a lot to do with everyday life, even the messy bits.

Jack-O'-Lantern

Omphalotus olearius

In the days before artificial lighting, the night was not a neutral thing: it was a threshold. A place of both terror and opportunity, where the rules of daylight loosened their grip and imagination was free to roam. Darkness was thought to be the hours of sprites and witches, fairies and phantoms, and anything that flickered or glimmered after sunset must have seemed particularly suspect. A world that suddenly began to glow on its own terms was not easily trusted. And yet, glow it did. From glow-worms in hedgerows to dancing fireflies, lightening balls to flowers that fluoresce in moonlight, nature has always had a fondness for theatrics. Some organisms, for instance, are biofluorescent, absorbing sunlight and re-emitting it later in altered hues. Others, more astonishing still, generate their own light from within. This process, known as bioluminescence, is a kind of quiet alchemy: squids and krill, octopuses and insects the world over flick on their own switches in the half-light. And, rather wonderfully, so do certain mushrooms.

Of the more than 120 species of fungi known to emit their own eerie glow, perhaps the most famous is the Jack-O'-Lantern mushroom. Native to central and southern Europe, particularly around its warmer coastlines, its Latin name reveals its preferred habitat: *Omphalotus olearius* translates loosely as 'belly button of the olive tree', a nod to its tendency to grow from the bases of Mediterranean hardwoods. An almost identical species, *Omphalotus illudens*, performs the same trick at higher latitudes, in northern Europe and North America alike.

Its folk name – Jack-O'-Lantern – shines a light on centuries of history. The name 'Jack' was long used as a derogatory term, implying someone who wasn't very good at something: fools were known as Jack a Dandies or Jack Fools, lazy men as Jack a Drognes or Jack a Dreams. Inferior animals were also tarred with the same brush – Jack Ass, Jack Rabbit, Jack Snipe, Jackdaw – but the word

could also imply a mischievous, knavish character: Jack the Lad or Jack Frost, for example. And to 'play the Jack' meant to lead someone astray. A Jack-O'-Lantern, or Jack a Lantern, was therefore someone or something that would trick others with their light, coaxing them off the path and into trouble. Shakespeare draws on this idea in *The Tempest*: 'you fairy, which you say is a harmless fairy, has done little better than played the Jack with us.'

The name Jack-O'-Lantern was not exclusive to mushrooms. It was also applied to *Ignis fatuus*, or will-o'-the-wisps, those ghostly flame-like flashes of light seen over marshes. Once thought to be malevolent sprites guiding travellers to watery graves, they are now understood to be caused by the spontaneous combustion of methane released by rotting vegetation. The Jack-O'-Lantern mushroom, however, is no less magical and unnerving. As night falls, these sublime organisms begin to glow blue-green around the gills, a trick accomplished by the interaction of a special enzyme with a light-emitting compound. Why mushrooms reserve their glowing secrets for the evening is not fully understood. Some scientists suggest the light attracts insects that help disperse spores; others believe it serves as a warning, quietly suggesting that a mushroom dinner might be a bad idea.

The idea of nature spirits luring the unwary into trouble with their glowing lanterns is ancient. The name 'foxfire' appears as early as the 15th century and referred to both *Ignis fatuus* and glowing mushrooms. The term is thought to come from the French *faux*, meaning false, although the fox itself has been associated with trickery since Anglo-Saxon times. For centuries, tagging our tailed friend onto a plant or fungus name – foxglove, foxbane, foxberry – was a cultural shorthand. It warned those who found it in the wild that this plant, or mushroom, wasn't quite to be trusted and belonged more to the land of fairies than the human realm.

Enoki

Flammulina filiformis

If you took a handful of wild enoki and compared them to a bunch of cultivated enoki, you'd be forgiven for thinking you were looking at two different species. Wild enoki grows in dense clusters on dead or dying trees. With its plush stem and wide rust-coloured cap, it is also known as the 'velvet foot' but, for all intents and purposes, looks very similar to many of its non-descript cousins.

For centuries, however, the enoki has also been cultivated in East Asia – perhaps even as far back as the 9th century CE – a process that transforms the rough wildling into a thing of delicate, ethereal beauty. Rather like forced rhubarb, which is grown in the dark to produce slim, pale, sweet stems, the enoki is subjected to a life in the gloom. Raised with very little light and under chilly, winter temperatures, the cultivated enoki becomes a ghostly, spectral white and grows long, spindly stems with almost no cap to speak of. Harvested in dense clusters, a handful of cultivated enoki looks like a bunch of crochet needles. And the mushroom has, indeed, knitted itself into centuries of mealtimes and medicine.

Its folkloric associations come from the mushroom's natural affinity with the cold; a fungus that thrived in winter, when the forests offered little else in terms of edible riches, earned the enoki a reputation for endurance, resilience and good health. Traditional Chinese Medicine and Japan's healing equivalent, Kampo, have both long valued the enoki as a remedy for fatigue and its ability to fortify a person's *qi* or vital energy, and is often prescribed to convalescing patients or the elderly. Korean traditional medical tradition similarly frames this fine filament of a mushroom – *paengi beoseot* – as a careful, mending thread after an unravelling illness or a difficult life transition.

FUNGI FAMILY

For centuries, botanists thought fungi were plants. And yet, they're closer – taxonomically speaking – to people than greenery. Like all animals, fungi don't photosynthesize; they absorb nutrients from their environment, typically secreting enzymes to digest food externally. Plants, by contrast, produce their own food via sunlight and photosynthesis.

But that's pretty much where the similarities end. In fact, fungi have a kingdom all of their own. Biologist have named about 150,000 species so far – from the mushrooms popping up on your lawn to the mould on your stale bread – but scientists suspect there are millions more still hiding in damp, wonderfully decaying corners of the world.

Some fungi grow mushrooms. Think of a mushroom as the fungus's fruit, like an apple on a tree. Its job is to release spores to spread the fungus far and wide. But the real hero is the mycelium – the underground network of tiny thread-like hyphae that burrow through the substrate like the roots of a plant. Some mycelium webs can stretch metres, even miles, under the soil or through decomposing material, seeking nutrients as they go.

Not all fungi, however, produce mushrooms. Yeasts are tiny, single-celled fungi with no threads and no mushrooms, but do an excellent job adding rise to bread or fizz to booze. Moulds, on the other hand, form plenty of fuzzy mycelium but don't put out any mushrooms. Then there are lichens – the ultimate team players. A lichen is a fungus plus one or more algae, living together as one happy unit, each getting something beneficial from the other.

And what about 'toadstools', that elusive of all categories? Strictly speaking, there's no such scientific group of fungi, but folklore and literature know better. Not all fungi are mushrooms, and not all mushrooms are toadstools. But all deadly or poisonous mushrooms are toadstools, from a long-standing and ancient belief that toads were terrifically toxic and produced by decaying matter or slimy water. Indeed, the word toadstool appears as early as the 13th century (as tadstoles) and lots of variations have leaped off the pages over the centuries. Toxic mushrooms have also been known as toad's meat, toad's cheese, toad's caps, toad's hats and frog stools.

Honey Fungus

Armillaria ostoyae

Around 8,000 years ago, the world was undergoing tremendous stress. The last Ice Age had ended, and the planet was in the process of heating up. Out of this climatic shift, life was transforming: sea levels rose, cutting off landmasses and creating islands such as Britain; hunter-gatherers across the world, often reluctantly, were turning to farming to cope with environmental uncertainty; and, over in Oregon's Blue Mountains, the spore of one lonely mushroom found its perfect home on a forest floor. That lone organism grew just a few feet every year, slowly spreading its tough mycelium far and wide through the soil. From just one spore, this single Honey Fungus now covers over 2,300 acres of woodland and is thought to be one of the oldest and largest living organisms on Earth.

When it fruits every autumn and pushes up mushrooms, *Armillaria ostoyae* does a good impression of your average, mild-mannered fungus. But looks can deceive. Beneath the surface, this extraordinary organism forces its way, relentlessly, through the soil, attacking the roots of vulnerable trees and cutting off their nutrients until they perish. While many fungi are simply useful decomposers of dead material, the Honey Fungus is both recycler and agent of destruction. For some, it is a simple villain; for others, a hero of the ecosystem – a valuable culler of frazzled, weak trees to make way for healthier new growth. Whatever you think of *Armillaria ostoyae*, it is, above all else, a survivor. And as the climate once again warms, many life forms – including humans – are being placed under renewed stress. Scientists suspect that the Honey Fungus, with its hidden, unstoppable underground life, may be one of the few organisms able to withstand the uncertainties ahead.

Blood-Red Webcap

Cortinarius sanguineus

When imagining the clothes our ancestors wore, it's easy to picture a world of beige flax and dull wool greys. Yet for textile historians, the world of ancient fabrics is gradually revealing itself to be dyed in full technicolour, rich with hues that once signalled wealth, identity and craftsmanship. Take the Vikings, for example – a broad term for the various Nordic seafaring communities who traded, farmed and pillaged across Europe, the British Isles, and even as far as the Middle East and America between the 8th and 11th centuries. Not only were they formidable warriors, the Vikings were also masters of textiles and dyeing, producing a wide spectrum of vibrant colours from both local plants and exotic pigments acquired through far-reaching trade networks.

Scandinavia's own landscape offered many natural options. The leaves of the woad plant yielded deep blues, hedge bedstraw produced vivid oranges, and walnut shells gave rich browns. Yellows were relatively easy and abundant, made from heather, mayweed, tansy or weld. Red, however, was far more challenging. High-status Viking garments suggest that the elite imported madder root and kermes (a red dye derived from Mediterranean insects) or acquired foreign, pre-dyed fabrics, making red a colour that symbolized wealth, prestige and power.

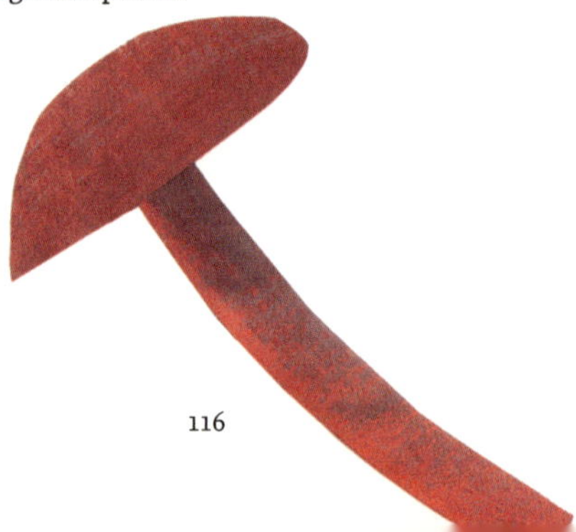

But all was not lost for those who couldn't access such exotic dyes. In the mossy, old-growth fir forests of northern Europe grows a sumptuous burgundy fungus – the Blood Red Webcap. This mushroom produces fabric dyes in an overflowing punnet of berry reds, raspberry pinks and plum purples, depending on the age of the specimen. Its colours are strikingly fruity, rarely matched by plants, and they take exceptionally well to wool; the warm, practical fabric of the cold northern climate. Today, modern dyers across Scandinavia, Europe and North America have rediscovered this vivid fungus, weaving the Blood Red Webcap back into the rich cloth of textile history.

Quinine Conk

Laricifomes officinalis

There is perhaps no mushroom whose medicinal reputation has travelled so far through time. The vast, hoof-shaped bracket fungus known as Quinine Conk was familiar to the Greek physician Dioscorides as *Agarikon*, a remedy of almost heroic ambition. For pains of the stomach, he advised chewing the bitter root and swallowing it dry; for tuberculosis, it was steeped in raisin wine; for the spleen, honey and vinegar softened its edge. It was medicine for a world in which the boundaries between food, cure and faith were still entirely blurred.

Fifteen centuries later, John Gerard was still recommending 'Agarick' for 'the shortnesse of breath, called Asthma, the inveterate cough of the lungs, [...] consumption and those that spet blood', its authority undiminished by the passage of time. The fungus earned its common name during the 19th century from its bitterness – reminiscent of quinine bark – and the assumption that it shared the same potent chemistry against fevers. It does not, in fact, contain a scrap of the stuff, but modern research has revealed something no less miraculous: strong anti-viral properties that go some way towards explaining its long medical career.

For many indigenous peoples of North America and Canada, the Quinine Conk was more than a medicine. The Squamish people of British Columbia, for example, displayed them in their homes to protect the occupants from malicious ill-wishers. Among the Alberta Cree, it was dried, powdered and applied to frostbite; larger conks were carved into figures and used in shamanic ritual, or placed as guardians at graves to protect a shaman in the afterlife. The Haida of the Pacific Northwest Coast also carved the fungus into effigies and used it to treat infections, swelling and bleeding – a living material that moved easily between the practical and the sacred. When Europeans crossed the Atlantic in the early 17th century, the Quinine Conk once again proved its worth. The devastating disease of smallpox arrived with the white colonists and this time-honoured healing mushroom offered indigenous communities one of their only chances of survival.

Lion's Mane

Hericium erinaceus

Beauty is in the eye of the beholder. For many people, *Hericium erinaceus* is about as ravishing a mushroom as you could wish for. With its creamy-white, cushioned body and cascading, hair-like fronds, it looks less like a fungus and more like a living creature. Cultures across Europe, Asia and North America have long been struck by this animated, lifelike appearance, and have named it accordingly. In Latin, it's *erinaceus* – the hedgehog – while in English it is the Lion's Mane. In German it becomes the Hedgehog's Spiky Beard; in France, the White Pom Pom. In China, where it is eaten with relish, it is known as Monkey-head. The Dutch see a Wig Mushroom, while in Japan it is variously called the Shaggy-robed Monk or Mountain Priest. Only in Korea is the name less poetic: the Roe Deer's Backside, inspired by its fluffy, rump-like shape.

Everyone sees something different in *Hericium erinaceus*, but all acknowledge its astonishing looks. Yet the Lion's Mane is far more than a visual curiosity. It is typically found growing wild on hardwood trees, across temperate forests in Asia, Europe and North America. Its favourite haunts are beech, oak and maple, and spotting one in the wild can feel like stumbling upon a forest oddity, an albino mammal perhaps or a discarded hairpiece.

Beyond its appearance and habitat, the most intriguing thing about Lion's Mane is its long-standing use in folk medicine. For centuries it has been a staple of Traditional Chinese Medicine, prescribed as a tonic for the heart, liver, lungs, spleen and kidneys, and valued for its ability to improve memory. What traditional uses have long suggested, modern medicine is only beginning to confirm. Contemporary studies indicate that Lion's Mane may support nerve growth, improve mild cognitive impairment and potentially help slow the progression of neurodegenerative disorders – a striking example of science catching up with folk knowledge.

Liberty Cap

Psilocybe semilanceata

On a mild October morning in 1799, a father, mother and their four children settled down to an early morning picnic in Green Park, London. The family often enjoyed open-air breakfasts of freshly gathered fungi, cooked over a fire in a small saucepan, all washed down with a steaming mug of tea.

This particularly morning, however, things soon took a sinister turn. About an hour after finishing his food, Edward – their eight-year-old son – started to laugh loudly and manically. Attempts to quieten him failed and the boy suddenly stopped his hysterics, became horribly dizzy and then fell into a deep and unawakenable stupor. His father soon succumbed to the same fate, followed by their ten-year-old daughter Charlotte, who had also enjoyed a large helping. Only Harriet, 12, and Martha, 18 – who had only eaten a small amount of the mushroom feast – got away with intense, but brief, attacks of vertigo.

Although some of the attending doctor's 'cures' were dubious – including drinking smelling salts, injecting his patients with a mysterious substance and rubbing their feet briskly with hot flannels – thanks to his quick interventions, and a large helping of providence, none of the family died. The physician later sent examples of the small and innocuous-looking mushrooms to a professor of botany at Oxford University for identification.

The suspect fungus turned out to be *Psilocybe semilanceata*, an incredibly common and ferociously potent psychoactive mushroom whose effects are much like LSD. Many now know it as 'magic mushrooms' but it has an older, and much more fascinating folk name: the Liberty Cap. In Roman times, when slaves were given their freedom, they were allowed to wear a simple, woollen hat called a *pileus*. Its name comes from the word for 'felt' and demonstrated to all those who saw it that a person was of a lowly status but had been granted their liberty. Even after the Empire fell, for centuries the symbol of a *pileus* remained as a marker of hard-won freedom across Europe and North America and was often depicted propped at the top of a long pole, like a battle standard. During the American Revolution, the French Revolution, and radical protests in Britain – such as the Peterloo Massacre in 1819 – the 'liberty cap' became the rallying symbol of resistance.

The first person to notice the similarity between the *pileus* and the hallucinogenic mushroom, and commit it to paper, was a little-known shoemaker-turned-nature writer called James Woodhouse. In his 1803 published work *Autumn and the Redbreast, an Ode*, Woodhouse revelled in the different shapes of native mushrooms and spotted 'Freedom's cap' among the woodland duff:

'Shap'd like War's shield, or Prelate's crown –
Like Freedom's cap, or Friar's cowl,
Or China's bright inverted bowl.'

The name clearly struck a chord and, just a few years later in 1812, the famous poets Samuel Taylor Coleridge and Robert Southey wrote about the 'Cap of Liberty' mushroom in their book of miscellany called *Omniana*: 'there is a common fungus, which so exactly represents the pole and cap of Liberty, that it seems offered by Nature herself as the appropriate emblem of Gallic Republicanism.'

More than a century later, in the 1950s, science finally identified and isolated the main psychoactive chemical in 'trippy' mushrooms – *psilocybin* – and began looking for it in other fungi. The Liberty Cap, it turns out, contains a hefty dose and soon became a poster child for the psychedelic movement and hippy counterculture. In 2005, the British government moved to ban 'magic mushrooms' outright, classifying them as a Class A substance alongside heroin and cocaine. Possession now carries the possibility of a seven-year prison sentence and an unlimited fine, while intent to supply can result in far harsher penalties. Whether foraged from a damp meadow or deliberately sought out, the consequences of owning this tiny fungus are severe. There is, perhaps, no greater irony than this: that a mushroom named for freedom, inspired by centuries of resistance and revolution, is now something that can very easily deprive you of it.

Ergot

Claviceps purpurea

Bog bodies are among the most intriguing – and gruesome – discoveries in archaeology: people ceremonially killed and deposited in peat bogs across northern Europe. Most date to the Iron Age, roughly 800 BCE to the 1st century CE, yet who they were and why they were so brutally sacrificed remains unknown. What unites these bodies, nevertheless, is the nature of their deaths. They were not simply killed, but subjected to prolonged, ritualized violence far beyond what was necessary. In some cases, analysis of stomach contents shows that they were fed baffling substances shortly before death, including mistletoe, moss – and ergot.

Ergot is a tiny black fungus that infects cereal crops, especially rye. Its curved shape gave rise to its folk names: 'horn feed' or ergot, from the Old French *argot*, meaning cock's spur. Ingested, this minuscule fungus produces two devastating effects. The first attacks the body: intense burning sensations in the hands, feet and limbs, followed by seizures and gangrene. In the Middle Ages this was known as St Anthony's Fire, named for the monastic order that treated its sufferers; today it is recognized as ergotism. Medieval physicians and midwives also discovered that smaller doses caused uterine contractions, and used it, haphazardly, to hasten labour or induce abortions.

Ergot's second effect is neurological. It can cause psychosis, violent headaches, delirium and vivid hallucinations. History records numerous outbreaks of ergotism, often among peasant communities, when damp winters followed by warm springs allowed infected grain to enter the food supply. These episodes were accidental. To deliberately administer ergot, however, suggests a far darker intent. We may never know why some of Europe's bog bodies were fed ergot-laced grain before their deaths. But its fiery pain and hallucinatory visions raise a disturbing possibility: were altered states of terror and madness themselves part of the grisly ritual?

Tippler's Bane

Coprinopsis atramentaria

Most mushrooms divide into one of three categories: delicious; not particularly edible but won't do you any harm; and the downright toxic. Just occasionally, however, you stumble upon a fourth type of fungi – one that is hiding a terrible trick up its sleeve. Eaten by itself, the Common Inkcap is, in theory, perfectly palatable if not a little dull. But – and this is a significant caveat – anyone who tries to make a meal of this innocuous-seeming mushroom and decides to wash it down with a cheeky glass of red or a tot of brandy will find themselves in grave danger.

The folk name for the Common Inkcap is the 'Tippler's Bane', literally the drinkers' ruin, a moniker that reveals at least some people in the countryside have long known of the mushroom's dark side. Rural herbal lore often grouped the inkcap with plants that punished excess, warning that nature always found ways of correcting human overindulgence. It seems science, nevertheless, only discovered the fungus' fickle nature later when, in the early 1900s, French mycologists first recorded mushroom-alcohol interactions and thought to take a closer look.

As mushrooms go, it is no great looker, nor is it particularly rare. Common across the northern hemisphere, and even parts of Australia and South Africa, it's a fungus especially fond of dungy soil, rotting wood and disturbed ground (its Latin name – *Coprinopsis* – comes from the ancient Greek *Kópros*, meaning excrement). Most people would also wander past a clump of Tippler's Banes between spring and autumn and barely give these miniature grey-brown specimens a second glance. Young, they resemble small, partially closed parasols but soon open fully, like an umbrella pulled flat in a gale, and begin to turn a sinister, raven black. This transformation is not merely aesthetic: inkcaps undergo a process called *deliquescence*, in which the gills digest themselves, collapsing into a slick, inky liquid.

Despite its unremarkable appearance, the Tippler's Bane dazzles when it comes to chemical composition. The fungus contains a compound called 'coprine', a name inspired by the plant's Latin botanical binomen and one that stops the human body being able to metabolize alcohol. A mouthful of Tippler's Bane, followed by a swing of booze, causes a sudden build-up of acetaldehyde, a toxic

byproduct that floods the body. Medical reports of 'coprinus syndrome' often begin with patients enjoying a rather delectable plate of sautéed fungi, paired with a crisp wine, before finding themselves with a thundering headache, vomiting, facial flushing, tachycardia and shortness of breath. These symptoms usually last only a few hours, however, before abating. Yet the coprine remains in the system for up to three days, waiting patiently to trigger another reaction should the sufferer give in to temptation and take another sip of anything alcoholic. Even more interestingly, the mushroom appears more potent when cooked, a cruel twist for anyone who prides themselves on culinary caution when it comes to foraging.

Knowledge of the Tippler's Bane's secret wasn't always widespread. In George Francis Atkinson's 1899 learned book, *Studies and Illustrations of Mushrooms*, the eminent mycologist related a story where he chastized two young friends that they were missing out by not eating the Common Inkcap and only sticking to Field Mushrooms (*Agaricus campestris*). Atkinson, in his book, went on to heartily recommend the gently nutty flavour of the Tippler's Bane, not raw, but 'properly prepared for table'. One can only hope that Atkinson and his companions decided to forgo the after-dinner port that night.

King Alfred's Cakes

Daldinia concentrica

During the 19th century, British academics developed a sudden enthusiasm for local folklore. As rapid industrialization transformed the countryside, many long-held rural traditions seemed in danger of vanishing in the smoke and steam. In response, a new generation of 'folklorists' hurried to record what they could. One especially fertile area was folk medicine; or, as a journalist in an 1890 edition of *The Leeds Mercury* put it, people's 'accompaniments of herbs, natural poultices, and strange diets of slugs, snails and mice.'

Belief in the curative power of objects was widespread. Simply carrying certain items on the body, especially in the pocket, was thought to draw pain away and encourage healing. A potato might ease rheumatism, for instance, or a puffball fungus could stop bleeding. A hairy worm cured whooping cough while a live frog treated thrush. For muscle pain, however, nothing rivalled the 'cramp ball' – a hard, coal-lump of a fungus known scientifically as *Daldinia concentrica*. This practice belonged to a magical view of healing, in which illness could be transferred into a natural object through its proximity. Once the pain had been absorbed, the object was discarded, taking the ailment with it.

Daldinia concentrica is also an excellent tinder fungus, holding an ember for hours, but it is perhaps best known for its other curious folk name, King Alfred's Cakes. The story of Alfred the Great burning bread while hiding from Viking raiders was first recorded in the 12th century, describing an incident from the late 870s. Distracted, Alfred allowed loaves to burn in a peasant's oven and was roundly scalded by the housewife – a reminder that even kings require instruction. Exactly when *Daldinia concentrica* was linked to this tale is unknown and is probably a Victorian or even early 20th-century invention. Facts never got in the way of a ripping yarn, however, especially in folklore.

Vegetable Cicada

Isaria sinclairii

Occasionally, you find a fungus that behaves in a way that's thrillingly awful. Say hello to the Vegetable Cicada, an organism that ruthlessly infects the bodies of baby insects and turns them into living zombies. Cicadas – those noisy, summer singers – spend most of their lives not flying around, but as wingless nymphs in the soil. Underground, they potter around for years, burrowing tunnels and drinking the sap of tree roots, before finally moulting and emerging as winged adults.

The Vegetable Cicada is a parasitic fungus that also lives under the surface. It finds its way inside these tender, young larvae-like insects and, once infected, a baby cicada soon loses control of its body and becomes a prisoner to the will of its invader. At first, the nymph starts to slow down and move sluggishly, its skin slowly changing to a ghostly white. The baby cicada is then directed, by the fungus, to wriggle into an ideal place for spore dispersal before finally succumbing. Once dead, the nymph is slowly gutted from the inside out – its own body mummified by fine threads of hyphae. When the warm weather arrives, the fungus finally makes a break for the soil surface, bursting out of the cicada's head and emerging out of the woodland floor like a clump of tiny, dusty white cotton buds.

The fungus also produces a magical compound called myriocin. In early studies it was shown to have formidable immunosuppressant activity, but was just too powerful and toxic to use in humans. So the medical world copied myriocin and produced a synthetic version – fingolimod – a treatment that is helping multiple sclerosis sufferers all over the world. Although the life cycle of the Vegetable Cicada has a whiff of horror movie about it, the end results are strangely beautiful.

Veiled Lady

Phallus indusiatus

If there were a prize for the best-dressed mushroom, the Veiled Lady would claim it. *Phallus indusiatus* dons the most exquisite petticoat or hooped crinoline, cutting an elegant figure in the moist woodlands and gardens of tropical regions. Lacey, delicate and breathtakingly aesthetic, it is perpetually poised – mid-swish – in a white, billowing underskirt. Indeed, the second part of its Latin name, *indusiatus*, literally means 'wearing underclothing'. It's a racy epithet for the French botanist Étienne Pierre Ventenat to have bestowed on the mushroom in 1798, but what's a little underwear reference when the genus name translates to 'erect penis'?

Sartorial elegance aside, it is quite the malodorous mushroom. Like all members of the stinkhorn family, its pollinator allure lies in a putrid, dead-flesh perfume, wafting from its slimy, olive-brown hat and intensifying with age. The Veiled Lady sends curiously mixed signals to humankind: enchanting beauty coupled with an abominable odour has left many cultures uncertain of its true nature. The forest-dwelling Sholaga people of southern India, expert mycologists, choose to leave it untouched despite the mushroom's edibility. In Madagascar, it is also regarded as highly poisonous. Over in the Democratic Republic of the Congo, the Veiled Lady was traditionally burned and its ashes rubbed onto the skin to treat paralysis, and yet also wielded in folk magic as a curse to disrupt friendships and lovers.

Asia harbours no such mycophobia. There, the pharmaceutical industry is discovering that the *Phallus indusiatus* contains some very promising compounds for 'cosmeceuticals' – beauty products with bioactive ingredients. The Veiled Lady, or Bamboo Mushroom, as it is better known, is also considered a delicacy, historically reserved for opulent banquets. Today, China cultivates this exceptional fungus on an impressive scale, bringing a taste of high fashion – and haute cuisine – to stir-fries and soups.

THE GOOD FORAGING GUIDE

KEEPING SAFE

TAKE A FUNGI COURSE. If you can, learn alongside an experienced forager. A walk in the woods with a knowledgeable guide or a hands-on foraging course is the best way to start.

REFER TO A BOOK. Carry an excellent field guide, one with clear colour photographs and good descriptions, and make it your foraging bible. Roger Phillip's *Mushrooms*, for example, is the gold-standard for UK fungi.

DON'T RISK IT. Only eat mushrooms you can identify with absolute cast-iron confidence. When it comes to fungi, over caution can save your life.

AVOID POLLUTION. Ditch gathering mushrooms from roadsides, industrial land, or polluted ground. They can sometimes have picked up noxious substances.

ESSENTIAL KIT

GRAB A BASKET. Use a woven carrier for your finds, rather than a plastic bag. A basket keeps your mushrooms fresh and unsquashed, and also allows spores to drift back to the forest floor as you walk.

BRING A FORAGING KNIFE. Slice mushrooms cleanly at the base with a small foraging knife rather than roughly pulling them up. This helps protect the delicate mycelium beneath the soil.

USE A SOFT MUSHROOM BRUSH. These gently remove any soil debris and leaf litter, without using water, which helps to preserve a mushroom's flavour and texture.

BE ORGANIZED. Keep different kinds of fungi separate in your basket so that, if small pieces snap off in transit, they can still be identified when you get back home.

BRING A CAMERA OR SMARTPHONE. Snapping your finds as you spot them in the wild can help with later study and identification.

KEEP A MUSHROOM DIARY. Make notes of where and when you found a particular species. Over time, these scribbles become a valuable, larger map of the seasons and your local woodland.

FORAGING BASICS

PAY ATTENTION TO CONTEXT. Notice the places and substrates mushrooms prefer. Some favour old oak or birch, for instance, while others love pine forests or meadow clearings.

WATCH THE WEATHER. The best time to look is often a few days after rain, when the earth is moist and the air mild. Some fungi will only emerge after a shower.

GET YOUR EYE IN. Like hunting for pebbles on a beach, it's good to train your eyes to the forest floor. Mushrooms can be hidden beneath leaves, nestled between moss, or tucked against fallen branches.

LOOK FOR CLUSTERS. When you find one mushroom, pause and look around. They often grow in family gatherings rather than go it alone.

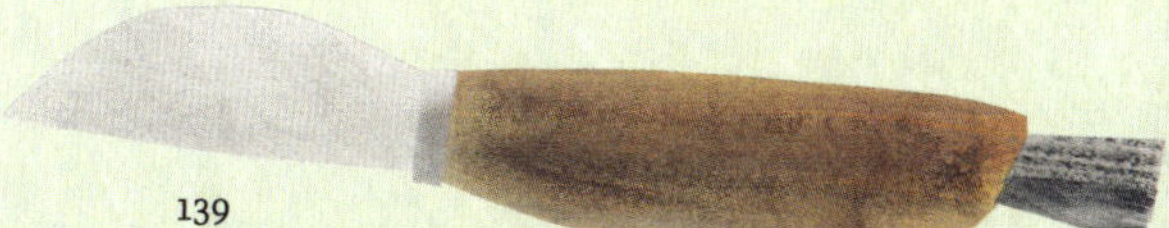

GOOD PRACTICE

START SMALL. Begin slowly and learn to confidently identify a handful of mushrooms rather than hunting for hundreds of different kinds.

LOOK CLOSELY. Pay attention to the smallest details: the shape of the cap, the colour of the gills, the contours of the stem, and the distinct smell. These subtle clues are the best way to distinguish between similar-looking species.

PICK SPARINGLY. Harvest modestly – just enough for the kitchen. The woodland can be a generous place, but it should never be stripped bare.

BE SELECTIVE. Let the newest, baby mushrooms develop and the oldest finish their work of releasing spores. Both play their part in the life cycle of the woods. Old mushrooms also often develop an unpleasant taste.

BE MINDFUL. Tread carefully. Moss, leaf litter, and decaying wood are home to countless plants, insects and small mammals, as well as other living things – keep disturbance to a minimum. Don't pick in areas of Special Scientific Interest (SSSIs).

DON'T PICK TO SELL. Most countries' foraging rules allow mushroom picking for personal use only, not for commercial sale (which usually needs a licence). Also, there may be species that are protected – in the UK, for example, Lion's Mane, Bearded Tooth and certain Waxcaps are not allowed to be picked because of their rarity.

Also by Sally Coulthard

Bird Lore

Even today, birds are hugely symbolic. From songbirds to seagulls, nightingales to morning larks, we still use the cultural language of the avian world. Doves bring peace, storks carry newborn babies. Robins mark Christmas, hens Easter. Blackbirds and thrushes herald the dawn, owls welcome in the night.

Even in our modern, rational world, magpies and peacocks still invite superstitions. Covering well-loved avian species from across the world, she glides through their folklore, legends and symbolism. From common sayings to strange superstitions, ancient beliefs to modern celebrations, this gloriously illustrated and international guide reveals that *Bird Lore* is as alive and vibrant as it ever was.

Tree Lore

Trees have deep roots in our folklore and mythology. Almost everything about them – leaves, branches, roots and fruits – has provided a metaphor for our human needs and desires. From strength to resilience, shelter to sustenance, we have taken each species' unique qualities and turned them into fables, sayings and symbolism.

Sally Coulthard branches out into the world folklore and finds an enchanted place where nature and legend intertwine. Whether it's lovers' trysting trees, festive firs, or boughs of peace, discover 50 of the world's most iconic species, their intriguing myths and magical associations.

Floriography

For thousands of years, flowers have served as a language of emotion. In this new book, Sally Coulthard, explores floriography, revealing how this secret language still shapes meaning across cultures, from ancient myth to hedgerow folklore.

Featuring 50 well-loved plants, from peonies to sweet peas, ivy to irises, *Floriography* is a beautifully illustrated guide through the history, legend, literature and symbolism of flowers, showing how their meanings continue to bloom today.

Acknowledgements

When I was about seven years old, I bought myself a copy of Roger Phillip's *Mushrooms* with my pocket money, a book so plump and gloriously photographed that I instantly fell in love with fungi. I also brought the book to show-and-tell at my rural primary school, to share my newfound passion. When the class had finally stopped laughing, and wiping the tears from their cheeks, I realized that perhaps I might have to enjoy this passion in secret, at least for a while. And so, over 40 years later it feels like I have finally found my crowd; a group of amazing women at Quadrille – Harriet, Gemma, Clover – who find as much pleasure as I do in the strange beauty of mushrooms. Thank you – you've made a childhood dream come true.

About the Author

Sally Coulthard is a bestselling nature writer and rural historian, captivated by the countryside and its people, plants and animals. She lives on a Yorkshire smallholding, where she divides her time between writing, growing food, and looking after a rag-tag assortment of feathered, furred and woolly companions.

Quadrille, Penguin Random House UK,
One Embassy Gardens, 8 Viaduct Gardens,
London SW11 7BW

Quadrille Publishing Limited is part of the Penguin Random House group of companies whose addresses can be found at global.penguinrandomhouse.com

Published by Quadrille in 2026

www.penguin.co.uk

A CIP catalogue record for this book is available from the British Library

ISBN 978 1 80712 011 5

10 9 8 7 6 5 4 3 2 1

Managing Director Sarah Lavelle
Editorial Director Harriet Butt
Senior Designer Gemma Hayden
Illustrator Clover Robin
Production Director Stephen Lang
Senior Production Controller Martina Georgieva

Colour reproduction by p2d

Printed in China by C&C Offset Printing Co., Ltd.

The authorized representative in the EEA is Penguin Random House Ireland, Morrison Chambers, 32 Nassau Street, Dublin D02 YH68.

Penguin Random House is committed to a sustainable future for our business, our readers and our planet. This book is made from Forest Stewardship Council® certified paper.

MIX
Paper
FSC® C018179